Mental Health in Historical Perspective

Covering all historical periods and geographical contexts, the series explores how mental illness has been understood, experienced, diagnosed, treated and contested. It will publish works that engage actively with contemporary debates related to mental health and, as such, will be of interest not only to historians, but also mental health professionals, patients and policy makers. With its focus on mental health, rather than just psychiatry, the series will endeavour to provide more patient-centred histories. Although this has long been an aim of health historians, it has not been realised, and this series aims to change that.

The scope of the series is kept as broad as possible to attract good quality proposals about all aspects of the history of mental health from all periods. The series emphasises interdisciplinary approaches to the field of study, and encourages short titles, longer works, collections, and titles which stretch the boundaries of academic publishing in new ways.

Despo Kritsotaki

A Social History of Mental Health Care in Greece, c. 1900–1920

Hospitals, Patients, and Society

palgrave
macmillan

Despo Kritsotaki
Modern Greek History Research
Centre
Academy of Athens
Athens, Greece

ISSN 2634-6036 ISSN 2634-6044 (electronic)
Mental Health in Historical Perspective
ISBN 978-3-032-05763-1 ISBN 978-3-032-05764-8 (eBook)
https://doi.org/10.1007/978-3-032-05764-8

This Palgrave Macmillan imprint is published by the registered company Springer Nature Switzerland AG
The registered company address is: Gewerbestrasse 11, 6330 Cham, Switzerland

Το Γιώργος for being there from the very beginning

Prologue

Different people with different stories: Sophia P. in a "sub-comatose condition";[1] Armenia S., of "limited intellect", unable to take care of herself;[2] Iakovos P., an alcoholic who turned aggressive;[3] Triantafyllos T., stubborn, irritable, impulsive, violent and "daunted by no one";[4] Vasileios G., a public servant, having "symptoms of neurasthenia" due to not getting a promotion.[5] All of them were hospitalised in the early twentieth century in the Eginitio or the Dromokaitio, two of the main Greek mental hospitals of the time, both located in Athens. This book recounts their stories, as well as the stories of many others who spent a—long or short—part of their lives at one of these institutions, and reconstitutes the multiple pathways to and the complex meanings of psychiatric hospitalisation in Greece during the early decades of the twentieth century.

By researching the place of psychiatric hospitals in Greek society, I argue that psychiatric hospitalisation was at the same time at the society's

[1] *Πρόχειραι Ιστορίαι Νόσου Εισαγομένων Ασθενών Β΄* [*Draft Illness Histories of Patients Admitted B΄*], archive of the Eginitio Hospital, 9.2, Athens, p. 313 (henceforth Eg/9.2).

[2] *Βιβλίον Ιστορικών [Case Histories Book]*, archive of the Dromokaitio Hospital, Athens (henceforth Dr/Case histories), 1920–1921, p. 1351.

[3] Eg/9.2, p. 80.

[4] Dr/Case histories, 1917, pp. 872–73.

[5] Dr/Case histories, 1919–1920, pp. 1213–14.

edges and centre. Hospitalisation sometimes meant long-term isolation and marginalisation, while other times it meant only temporary removal from family and friends, colleagues and neighbours. Moreover, while hospital admission was a difficult decision—mental hospitals seemed far away and hostile institutions—it simultaneously fulfilled an array of social functions, which were shaped by the ways that mental hospitals operated, but also by the beliefs, practices, and needs of the families and communities around them.

In order to highlight the multiple dimensions and functions of psychiatric hospitalisation, I turn to the issue of who was hospitalised, by whom, and for which reasons. I explore this issue from the perspective of the hospitals and their medical staff, as well as the patients and their private and public milieu. Along the trend that first emerged in the historiography of psychiatry some forty years ago (see Chap. 1), this study attempts to distance itself from an analysis that insists on abstract social, economic, and political forces. To this end, I delve into the archives of two hospitals to bring to the foreground the specific subjects—individuals, social actors, and social groups—who lived and/or worked in these hospitals or endeavoured to use them in various ways. In addition, I investigate the professional and lay, private and public discourses around the hospitals, which allow an insight into the meanings not just of hospitalisation but also more generally of mental illness. The key question here is in what ways social characteristics and roles shaped the definition and management of mental illness.

But, why the early decades of the twentieth century? Internationally, this was still a time before community care had developed, a time of predominately intramural psychiatry, as the nineteenth century had been. In Greece, it was actually a time of timid expansion of inpatient mental services. Next to the existing mental hospitals, namely the public psychiatric hospital of Corfu, founded in 1838 by the British administration of the Ionian Islands, and the Dromokaitio, the charitable hospital of Athens, founded in 1887, the Eginitio was founded in 1904. This was the first university clinic of psychiatry and neurology in Greece, and its establishment manifested the growing recognition of neurology-psychiatry (a unified specialty) in the country. In other words, the early twentieth century is particularly suited for a study of intramural mental health care in Greece, as outpatient services were still practically non-existent, but psychiatry had a more solid institutional foundation, with additional and more diverse services than in the nineteenth century.

This book is based on the Ph.D. thesis that I defended in December 2009 at the Department of History and Archaeology of the University of Crete. Tasos Sakellaropoulos and Vangelis Karamanolakis helped me to turn to this department and I thank them warmly as it provided the ideal framework for my research. For welcoming me as a doctoral researcher and for their continuous support, supervision, and assistance, I thank my supervisor Professor Christos Loukos, and the two other members of the committee, Efi Avdela, Professor of the Department of History and Archaeology of the University of Crete, and Dimitris Ploumpidis, Associate Professor of the Medical School of the University of Athens. I have worked creatively with them in the following years, and I hope to work with them many more times in the future. I owe special thanks to Ioannis Evdokimidis, who, together with Dimitris Ploumpidis, gave me the opportunity to research the archive of the Eginitio Hospital, which they managed to save and preserve. I also thank the board of directors of the Dromokaitio Hospital during the years 2005–2007, for allowing me to study the amazing archival material kept at the hospital since its foundation.

The transition from thesis to book was not easy. Many years passed, as I turned to new research immediately after completing the thesis. Without the persistence of my colleague and good friend, Dimitra Vassiliadou, I doubt I would have devoted the energy needed to return to this work that I had completed about fifteen years previously. However, eventually, this return appealed to me, and I thank Dimitra for urging me so convincingly to attempt it.

Finally, I am grateful to the anonymous reviewer for their comments, to the editors of the Mental Health in Historical Perspective series, and, in particular, to Matt Smith, for their support, and for helping this book reach a broader audience outside Greece.

Contents

List of Diagrams

List of Tables

CHAPTER 1

Introduction

Writing the History of Psychiatry

The history of psychiatry was first explored and narrated by the psychiatrists themselves in the nineteenth century. Many psychiatric textbooks included historical introductions in which the birth of psychiatry and the asylum were presented as a breakthrough in the treatment of mental illness. According to these accounts, which reflected the professional perspective and interests of their authors, psychiatry was a science based on humanitarianism, charity and modern treatment methods, seeing as it mission to replace the barbarism and superstitions that had prevailed until the end of the eighteenth century.[1] At least until the 1960s, the dominant voice in the history of psychiatry was that of psychiatrists, who largely followed the same humanistic and progressive interpretive schema as their nineteenth-century predecessors, viewing the development of the psychiatric specialty, institutions and state care for the mentally ill as signs of scientific and social progress. The central motif of this type of history was

[1] See indicatively Daniel Hack Tuke, *Chapters in the History of the Insane in the British Isles*, London: Kegan Paul, Trench, 1882 and Simon Apostolidis, *Αι Ψυχώσεις [The Psychoses]*, Athens: s.n., 1889, pp. 9–10. Even when their tone was mild, these stories responded to the psychiatric profession's need to project itself as a modern science. Nikolas Rose, *Inventing our Selves: Psychology, Power, and Personhood*, Cambridge: Cambridge University Press, 1996, pp. 41–43.

D. Kritsotaki, *A Social History of Mental Health Care in Greece, c. 1900–1920*, Mental Health in Historical Perspective,
https://doi.org/10.1007/978-3-032-05764-8_1

the search for the precursors of modern psychiatry. As a result, historical figures, ideas, and events were assessed by anachronistic criteria.[2]

The impetus for a different history of psychiatry was given in the 1960s when Michel Foucault's *History of Madness* and writings by psychiatrists and sociologists critical of psychiatry introduced two key concepts: power and social and cultural construction. In analysing the path to psychiatric asylums from the late Middle Ages to the nineteenth century, Foucault focused on the increasing centralisation of state power, which was expressed in the seventeenth century in the "great confinement" of marginalised and often vagrant individuals, such as the poor, the unemployed, criminals and the insane, in general institutions, typified by the general hospitals in France and the workhouses in England.[3] The French philosopher challenged psychiatric asylums as places of charity and protection and in later work presented them, along with schools, prisons, and reformatories, as institutions of discipline and instruments of the knowledge-power dichotomy.[4] Foucault's work, despite the criticism it has encountered, contributed to the understanding of madness as a changing social construct and to this day inspires many debates about mental illness and psychiatry.[5]

[2] Gregory Zilboorg and George Henry, *A History of Medical Psychology*, New York: W. W. Norton & Company, 1941; Franz Alexander and Sheldon Selesnick, *The History of Psychiatry: An Evaluation of Psychiatric Thought and Practice from Prehistoric Times to the Present*, London: Allen & Unwin, 1967. These writers were psychiatrists, but the same view was shared by the few historians who did study these issues. For example, Albert Deutsch, *The Mentally Ill in America: A History of their Care and Treatment from Colonial Times*, Garden City, N.Y.: Doubleday, Doran & Company, 1937 and Kathleen Jones, *Lunacy, Law and Conscience 1744–1845*, London: Routledge Kegan Paul, 1955. For a critical review of this approach, see Jacques Postel, *Éléments pour une Histoire de la Psychiatrie Occidentale*, Paris: L' Harmattan, 2007, pp. 31–39.

[3] Michel Foucault, *Madness and Civilization. A History of Insanity in the Age of Reason*, transl. Richard Howard, New York: Vintage, 1973, Chapter 3 (first edition in French in 1961).

[4] Michel Foucault, *Surveiller et Punir: Naissance de la Prison*, Paris: Gallimard, 1975.

[5] The volume edited by Arthur Still and Irving Velody (*Rewriting the History of Madness: Studies in Foucault's Histoire de la Folie*, London: Routledge, 1992) is indicative. To date, it is difficult to find a study of the history of mental illness or mental health sciences that does not mention Foucault. For a recent study that also analyses Foucault's relationship with psychiatry, see Camille Robcis, *Disalienation. Politics, Philosophy, and Radical Psychiatry in Postwar France*, Chicago: University of Chicago Press, 2021, especially pp. 107–42.

During the same period, psychiatrists, sociologists and historians were beginning to question the legitimacy of psychiatry and denounce it as a specialty with too much power and without the humanitarian mission it advocated. These critiques, some of which have been grouped under the term "antipsychiatry",[6] focused on the concepts of social construction, power and social control and talked about the "myth of mental illness",[7] schizophrenia as a healthy reaction to a mad society,[8] the brutal way in which asylums functioned as "total institutions"[9] and the asylum as a means of disciplining women specifically and exercising patriarchal power.[10] At the same time, the "labelling theory" was gaining popularity through the work of the sociologist Thomas Scheff, who argued that psychiatric diagnoses were given to people who did not conform to social norms.[11] These arguments, which were taken up by literary works and films,[12] and the revelation of scandals concerning the methods and quality of psychiatric care in large asylums, led to scepticism about the meanings of mental illness and the role of psychiatry, and fuelled initiatives in many countries to replace asylums with alternative, community-based structures.

Under these influences, historians and sociologists began from the 1970s to test Foucault's and antipsychiatry's interpretations of the history

[6] According to Peter Sedgwick (*Psycho Politics*, London: Pluto, 1982), the main representatives of the "antipsychiatry movement" were Michel Foucault, Thomas Szasz, Erving Goffman and Ronald Laing. However, these academics and professionals did not use the term "antipsychiatry", while Szasz explicitly denied this identification. Richard Vatz and Lee Weinberg, "The rhetorical paradigm in psychiatric history: Thomas Szasz and the myth of mental illness", in Mark Micale and Roy Porter (eds), *Discovering The History of Psychiatry*, New York: Oxford University Press, 1994, pp. 311–30.

[7] Thomas Szasz, *The Myth of Mental Illness: Foundations of a Theory of Personal Conduct*, New York: Harper & Row, 1974 (first edition in 1960).

[8] Ronald Laing, *The Politics of Experience and the Bird of Paradise*, Harmondsworth: Penguin, 1967 (first edition in 1964).

[9] Erving Goffman, *Asylums: Essays on the Social Situation of Mental Patients and other Inmates*, Garden City, N.Y.: Anchor Books, 1961.

[10] Phyllis Chesler, *Women and Madness*, Garden City, N.Y.: Doubleday, 1972.

[11] Gerald Grob, "The history of the asylum revisited: Personal reflections", in Micale and Porter (eds), *Discovering the History of Psychiatry*, pp. 260–81.

[12] Most notably, the novel by Ken Kesey, *One Flew over the Cuckoo's Nest*, first published in 1962, which was also turned into a popular film in 1972, directed by Miloš Forman and starring Jack Nicholson.

of psychiatry. The protagonist of this historiography was the psychiatric asylum, its origins, development and operation in different countries.[13] At the centre of the analysis were the bourgeoisie and capitalism, as well as the processes of urbanisation and industrialisation, which, by causing economic and social upheavals, were perceived as leading to changes in the handling of the mentally ill.[14] In the perspective of these authors, the driving forces behind psychiatric reform in the late eighteenth and nineteenth centuries were not humanitarianism and scientific progress, but the efforts of the authorities and the ruling classes to preserve social order by controlling the lower, marginal and "dangerous" social strata. This argument was reinforced by the emphasis that psychiatric legislation in many countries placed on the "dangerous" and destitute "lunatics" (see Chapter 3) and their confinement in psychiatric asylums. Finally, the social control argument was supported by studies of everyday life and the doctor-patient relationship in institutions, where life was governed by discipline and industriousness, in order to instil in the patients these

[13] On North America, see David Rothman, *The Discovery of the Asylum: Social Order and Disorder in the New Republic*, Boston: Little, Brown, 1971; on France, Guy Le Gaufey and Gérard Bleandonu, "Naissance des asiles d' aliénés (Auxerre-Paris)", *Annales*, 1975 (30), pp. 93–121; on England, Andrew Scull, *Museums of Madness: The Social Organization of Insanity in Nineteenth-Century England*, London: Allen Lane, 1979; on England, France and Germany, Klaus Dörner, *Bürger und Irre: Zur Sozialgeschichte und Wissenschaftssoziologie der Psychiatrie*, Frankfurt am Main: Fischer Taschenbuch Verlag, 1975 (first published in 1969); on Ireland, Mark Finnane, *Insanity and the Insane in Post-Famine Ireland*, London: Croom Helm, 1981. The emphasis on the asylum was not accidental, as modern psychiatry from its inception was inextricably linked to the "therapeutic asylum", as it was conceived in the late eighteenth century. Edward Shorter, *A History of Psychiatry: From the Era of the Asylum to the Age of Prozac*, New York: John Wiley, 1997, pp. 8–22. Being asylum-oriented, psychiatry followed the general trend in medicine: from the end of the eighteenth century, hospitals were playing an increasingly important role in medical practice and in the organisation of the medical profession. Lindsay Granshaw, "The rise of the modern hospital in Britain", in Andrew Wear (ed.), *Medicine in Society: Historical Essays*, Cambridge: Cambridge University Press, 1992, pp. 197–218.

[14] Typical of these interpretations was Scull's analysis, which located the emergence of the psychiatric asylum in England in the mid-eighteenth century, when the spread of the market economy brought about major social changes, such as the introduction of wage labour and the shift from a master-servant to an employer-employee relationship. These changes, Scull argued, restricted the capacity of families and communities to care for the ill and led to the latter's removal to special institutions. Andrew Scull, *Social Order/Mental Disorder: Anglo-American Psychiatry in Historical Perspective*, London: Routledge, 1989, pp. 213–38.

two basic bourgeois principles of the nineteenth century, and where the patients were placed in the position of the "child" to be cared for and instructed by the doctor "father".[15]

Other motives for the psychiatric reform have been traced to the professional interests of psychiatrists, who belonged to or sided with the bourgeoisie, in order to secure economic and social status.[16] It has been argued that psychiatrists saw mental illness as a lucrative field of professional activity and fought to keep it under their exclusive purview.[17] As a result, the lunatic asylum was sketched out as the main means of consolidating medical power over situations and people that had previously fallen under other jurisdictions, such as those of the family, the church, the workhouse or the prison authorities.[18]

From the late 1960s onwards, less radical critiques of traditional psychiatric historiography also emerged, mainly by psychiatrists, who focused more on methodological issues and tried to use historical methods, studying primary and original sources, to distinguish facts from myths in the history of psychiatry. On the basis of tangible evidence, they sought to avoid value judgements or at least to distinguish them from historical facts, and to take into account the general historical background.[19] While still largely focused on "important" psychiatrists, these studies moved away from the hagiography of psychiatry's great "heroes", examined hitherto unknown psychiatric texts and brought wider social, economic and political factors into the analysis.[20] However, these researchers did not necessarily subscribe to the notions of psychiatry as an agency of power

[15] Foucault, *Madness and Civilization*, pp. 253–54; Eric Engstrom, *Clinical Psychiatry in Imperial Germany: A History of Psychiatric Practice*, Ithaca: Cornell University Press, 2003, pp. 20–22.

[16] Jacques Postel, *Éléments pour une Histoire de la Psychiatrie Occidentale*, pp. 32–33.

[17] Andrew Scull, *The Most Solitary of Afflictions: Madness and Society in Britain, 1700–1900*, New Haven: Yale University Press, 1993, pp. 175–216.

[18] Andrew Scull, "The social history of psychiatry in the Victorian era", in Andrew Scull (ed.), *Madhouses, Mad-Doctors, and Madmen. The Social History of Psychiatry in the Victorian Era*, London: Athlone Press, 1981, pp. 5–32.

[19] Henri Ellenberger, "La conférence de Freud sur l' hysterie masculine" [1968], in Élizabeth Roudinesco (ed.), *Médecines de l' Âme: Essais d' Histoire de la Folie et des Guérisons Psychiques*, Paris: Fayard, 1995, pp. 207–25.

[20] Gladys Swain, *Le Sujet de la Folie: Naissance de la Psychiatrie*, Toulouse: Privat, 1977; Henri Ellenberger, *The Discovery of the Unconscious: The History and Evolution of Dynamic Psychiatry*, London: Fontana Press, 1994 (first published in 1970); Jacques Postel, *Genèse*

and of mental illness as a social construct, nor did they question the good intentions of most doctors in the institutions, which, they argued, admitted genuinely and seriously ill people, not just those who challenged capitalism, patriarchy and the "status quo" in general.[21]

Since the 1980s, research into the history of psychiatry has proliferated, and its themes, methods and theoretical and ideological orientations have been greatly enriched. A new revisionist historiography criticised the radical historiography of the 1960s and 1970s on the grounds that it attributed exclusively to the state the function of social control and the safeguarding of social order and described all social relations as relations of power and submission.[22] The appropriateness of the explanatory schema of social, state and medical control was questioned[23] and the insistence on the dilemma of humanitarianism/ scientific progress versus social control/ oppression was seen as obscuring the complexity of the motives, meanings and functions of psychiatry.[24] The link between psychiatric institutions and state authority was relativised as it became clear that asylums were being established in countries with different systems of governance,[25] even in countries where the state had neither the power nor the will to impose a centralised system of institutions, and, as a result, asylums were not always run by the state.[26] Furthermore, it was convincingly argued that the hegemony of psychiatrists was far from complete,

de la Psychiatrie: Les Premiers Écrits Psychiatriques de Philippe Pinel, Paris: Le Sycomore, 1981.

21 Gerald Grob, *Mental Institutions in America: Social Policy to 1875*, New York: Free Press, 1973.

22 Michael Ignatieff, "State, civil society and total institutions: A critique of recent social histories of punishment", in Stanley Cohen and Andrew Scull (eds), *Social Control and the State: Historical and Comparative Essays*, Oxford: Robertson, 1983, pp. 75–105.

23 William Frederick Bynum, Roy Porter and Michael Shepherd, "Introduction", in William Frederick Bynum, Roy Porter and Michael Shepherd (eds), *The Anatomy of Madness: Essays in the History of Psychiatry,* vol. 2, London: Tavistock, 1985, pp. 1–16.

24 David Rothman, "Social control: The uses and abuses of the concept in the history of incarceration", in Stanley Cohen and Andrew Scull (eds), *Social Control and the State*, pp. 106–17.

25 Shorter, *A History of Psychiatry*, pp. 6–17.

26 For example, in England up to the middle of the nineteenth century more than half of the institutions were private. Roy Porter, *Madness. A Brief History*, Oxford: Oxford University Press, 2002, pp. 92–100.

as they were largely dependent on state and local authorities,[27] competed with and were challenged by other professionals, such as neurologists and hypnotists, and were often unable to provide adequate explanations and treatments for mental illness.[28] Although, as an emerging profession in the nineteenth century, they sought to increase their clientele, income and professional and social prestige, they were unable to act in a coordinated way to do so, because they were a heterogeneous group, had little influence as an organised body and, at least until the end of the nineteenth century, had a low social status.[29]

As researchers downplayed the role of psychiatrists and state authorities, new factors were introduced to explain the emergence and expansion of psychiatric asylums: cultural and ideological factors, especially the scientific thought of the Enlightenment, which gave rise to the new therapeutic optimism,[30] but also religious and political factors, for example, in the case of England, the Anglican attack on popular religious enthusiasm and

[27] On the state's role, see Andrew Scull, Charlotte MacKenzie and Nicholas Hervey, *Masters of Bedlam: The Transformation of the Mad-Doctoring Trade*, Princeton, N.J.: Princeton University Press, 1996, p. 7, and on that of the local authorities, see Richard Adair, Bill Forsythe and Joseph Melling, "A danger to the public? Disposing of pauper lunatics in late-Victorian and Edwardian England: Plympton St. Mary Union and the Devon County Asylum, 1867–1914", *Medical History*, 42 (1998), pp. 1–25.

[28] Nancy Tomes, "The anatomy of madness: New directions in the history of psychiatry", *Social Studies of Science*, 17 (1987), pp. 358–70.

[29] German Berrios and Hugh Freeman, "Introduction", in German Berrios and Hugh Freeman (eds), *150 Years of British Psychiatry, 1841–1991*, London: Gaskell, 1991, pp. ix-xiv and Edward Renvoize, "The Association of Medical Officers of Asylums and the Hospitals for the Insane, the Medico-psychological Association and their presidents", in Berrios and Freeman (eds), *150 Years of British Psychiatry, 1841–1991*, pp. 29–78. More recent studies also question the omnipotence of psychiatry, highlighting its ambiguities and weaknesses. For example, Jessie Hewitt (*Institutionalizing Gender. Madness, the Family, and Psychiatric Power in Nineteenth-Century France,* Ithaca: Cornell University Press, 2020) demonstrates the inability of asylum psychiatrists "to solidify notions of gender difference" (p. 18).

[30] Edward Shorter points out that the new psychiatric asylums spread to countries with very different economic and social conditions and administrative systems, and thus rejects general explanations based on growing state centralisation or capitalism, both central concepts for Foucault and Scull. In contrast, the new ideas of the Enlightenment were international, disseminated through journals, books and exchanges between doctors, and thus fed an international medical movement with common goals. Shorter, *A History of Psychiatry*, pp. 6–17.

measures against exorcisms and all kinds of magical cures.[31] Moreover, under the influence of the new social history and medical sociology, which at the same time began to turn with greater interest to the experiences of patients,[32] historical research has focused on the mentally ill, their social milieu and the role it played in the decision to admit them to psychiatric institutions.[33] More recent studies have explored the complex interactions between asylums and the social, political, economic and cultural contexts in which they operate,[34] using the history of specific psychiatric institutions to illuminate broader facets of modernity.[35] Such approaches have challenged the image of the asylum as a socially isolated institution.

Indeed, despite the emphasis on asylums, the multitude of studies that have appeared over the last thirty-five years have addressed larger issues, such as the cultural meanings of mental illness, the relationship between medical and "lay", i.e. non-medical, discourses and practices around

[31] Michael MacDonald, *Mystical Bedlam: Madness, Anxiety, and Healing in Seventeenth-Century England*, Cambridge: Cambridge University Press, 1981. For a history of psychiatry that takes into account ideas, philosophical and religious, as well as political events and social conditions, see Jan Goldstein, *Console and Classify: The French Psychiatric Profession in the Nineteenth Century*, Cambridge: Cambridge University Press, 1987.

[32] Julia Lawton, "Lay experiences of health and illness: Past research and future agendas", *Sociology of Health & Illness*, 25 (2003), pp. 23–40.

[33] For instance, Nancy Tomes, *A Generous Confidence: Thomas Story Kirkbride and the Art of Asylum-keeping, 1840–1883*, Cambridge: Cambridge University Press, 1984; Mark Finnane, "Asylums, families and the state", *History Workshop Journal*, 20 (1985), pp. 134–48; Constance McGovern, "The community, the hospital, and the working-class patient: The multiple uses of asylum in nineteenth century America", *Pennsylvania History*, 54 (1987), pp. 17–33; David Wright, "Getting out of the asylum: Understanding the confinement of the insane in the nineteenth century", *Social History of Medicine*, 10 (1997), pp. 137–55; Françoise Jacob, "'Folie' et étude de cas. Jean Longuebeau, ouvrier cloutier, un délire à Toulouse, en 1836", in François-Olivier Touati (ed.), *Maladies, Médecines et Sociétés: Approches Historiques pour le Present: Actes du VIe Colloque d'Histoire au Present*, Paris: L' Harmattan, 1993, pp. 64–75; Akihito Suzuki, *Madness at Home: The Psychiatrist, the Patient, and the Family in England, 1820–1860*, Berkeley: University of California Press, 2006; H. Yumi Kim, *Madness in the Family: Women, Care, and Illness in Japan*, Oxford, New York: Oxford University Press, 2022.

[34] Katherine Ziff, David Thomas and Patricia Beamish, "Asylum and community: The Athens Lunatic Asylum in nineteenth-century Ohio», *History of Psychiatry*, 19 (2008), pp. 409–32; Anatole Le Bras, *Aliénés—Une histoire sociale de la folie au XIXe siècle*, Paris: CNRS Éditions, 2024.

[35] Joelle Abi-Rached, *Aṣfūriyyeh: A History of Madness, Modernity, and War in the Middle East*, Cambridge, MA: MIT Press, 2020.

mental illness, as well as the experiences and social profile of patients.[36] Moving from the history of psychiatry to the history of mental illness, the focus is on the ways in which mental illness has been viewed, represented and treated in different periods and societies.[37] Clinical records, as well as autobiographical texts and letters written by patients, are studied in an attempt to reconstruct the experiences, thoughts and hopes of patients, as well as the values and fears of their time, and thus to produce a social history of madness or a medical history "from below".[38] The question is not so much whether mental illness is a real illness, but rather to what extent its traits are influenced by social experience and the intellectual, cultural and social landscape.[39] Within this frame, different diseases are not treated as objective pathological entities—a medical-centred perspective that prevailed until the 1980s—but as complex cultural and social constructs.[40]

[36] See, indicatively, William Frederick Bynum, Roy Porter and Michael Shepherd, *The Anatomy of Madness: Essays in the History of Psychiatry,* 3 vols, London: Tavistock, 1985–88; Greg Eghigian (ed.), *The Routledge History of Madness and Mental Health*, New York: Routledge, 2017.

[37] Roy Porter, *A Social History of Madness: Stories of the Insane*, London: Weidenfeld and Nicolson, 1987, pp. 5–9; Andrew Scull, *Madness in Civilization: A Cultural History of Insanity from the Bible to Freud, from the Madhouse to Modern Medicine*, London: Thames & Hudson, 2015.

[38] The constitutional text of medical history "from below" is Roy Porter's, "The patient's view: Doing medical history from below", *Theory and Society*, 14 (1985), pp. 175–98. On mental illness and psychiatric history specifically, see, for example, Allan Beveridge, "Life in the asylum: Patients' letters from Morningside, 1873–1908", *History of Psychiatry*, 9 (1998), pp. 431–68; Robert Allan Houston, *Madness and Society in Eighteenth-Century Scotland*, Oxford: Clarendon Press, 2000, pp. 119–20; Hazel Morrison, "Constructing patient stories: 'Dynamic' case notes and clinical encounters at Glasgow's Gartnavel Mental Hospital, 1921–32", *Medical History* 60 (2016), pp. 67–86; Benoît Majerus, "Making sense of the 'chemical revolution'. Patients' voices on the introduction of neuroleptics in the 1950s, *Medical History* 60 (2016), pp. 54–66; Katherine Hodgkin, *Madness in Seventeenth-Century Autobiography*, Basingstoke: Palgrave Macmillan, 2007; Le Bras, *Aliénés*.

[39] Roy Porter, *Mind Forg'd Manacles: A History of Madness in England from the Restoration to the Regency*, London: Athlone, 1987, p. 33; Houston, *Madness and Society in Eighteenth-Century Scotland*, pp. 13–19.

[40] The cases of hysteria and neurasthenia, which have been the subject of new cultural histories, are typical. For example, Sander Gilman, Helen King, Roy Porter, G. S. Rousseau and Elaine Showalter (eds), *Hysteria beyond Freud*, Berkeley: University of California Press, 1993; Edward Shorter, *From Paralysis to Fatigue: A History of Psychosomatic Illness in the*

In recent years, the historiography of psychiatry has taken many intriguing turns, with new insights emerging on the role of psychiatry in colonial and post-colonial contexts,[41] the "transnationality" of mental health[42] and the dimensions of space, materiality and the body.[43] Yet, the most significant renewal of the historiography of psychiatry has been brought about by studies focusing on gender. In the 1970s, much of the history of female mental patients was fuelled by antipsychiatry and the interpretive schema of control. The "feminist revisionist school" in the history of psychiatry took two central positions. Firstly, it was proposed that nineteenth-century psychiatrists labelled female behaviour that was socially and culturally unacceptable as mental illnesses, thereby controlling women's social and especially sexual behaviour. Secondly, it was argued that women constituted the majority of patients admitted to psychiatric institutions, given that the concept of "appropriate" female behaviour was less permissive than that of "appropriate" male behaviour.[44] However, in

Modern Era, New York: Free Press, 1992; Marijke Gijswijt-Hofstra and Roy Porter (eds), *Cultures of Neurasthenia from Beard to the First World War*, Amsterdam: Rodopi, 2001.

41 Ernst Waltraud, "Histories of madness in South Asia", in Greg Eghigian (ed.), *The Routledge History of Madness and Mental Health*, London: Routledge, 2017, pp. 210–28; Sally Swartz, "Mad Africa" in Eghigian (ed.), *The Routledge History of Madness and Mental Health.*, pp. 229–44; Richard Keller, *Colonial Madness: Psychiatry in French North Africa*, Chicago: University of Chicago Press, 2007.

42 Wu Harry Yi-Jui, *Mad by the Millions: Mental Disorders and the Early Years of the World Health Organization*, Cambridge, MA: MIT Press, 2020.

43 Monika Ankele and Benoît Majerus (eds), *Material Cultures of Psychiatry*, Bielefeld: Transcript, 2020; Jennifer Wallis, *Investigating the Body in the Victorian Asylum: Doctors, Patients, and Practices*, London: Palgrave Macmillan, 2017.

44 For a review of the historiography on mental illness and women and the feminist approach of the history of psychiatry, see David Wright, "Delusions of gender? Lay identification and clinical diagnosis of insanity in Victorian England", in Jonathan Andrews and Anne Digby (eds), *Sex and Seclusion, Class and Custody: Perspectives on Gender and Class in the History of British and Irish Psychiatry*, Amsterdam: Rodopi, 2004, pp. 149–76; Alexandra Wallis, "Hysterical women: Moral treatment of female patients in the Fremantle Lunatic Asylum, 1858–1908", Ph.D. thesis, The University of Notre Dame Australia, 2020, introduction and chapter 1. A key work of the feminist revisionist school was *Women and Madness* by Phyllis Chesler, which argued that the "label" of mental illness is meant to regulate and uphold male and female roles. Regarding the unique connection between femininity and madness, see also, Elaine Showalter, *The Female Malady, Women, Madness and English Culture, 1830–1980*, New York: Pantheon Books, 1985 and Yannick Ripa, *Women and Madness: The Incarceration of Women in Nineteenth-Century France*, Cambridge: Polity Press, 1990 (first published in 1986).

the context of new research in the history of psychiatry since the 1980s, both the view of female patients as victims of patriarchal society and male-dominated psychiatry and the quantitative predominance of women over men in nineteenth-century psychiatric institutions have been subjected to critical scrutiny.[45] The multiplicity of meanings attributed to the asylum for both men and women has been a prominent focus of analysis. Medical texts have been employed to investigate the meanings of femininity and the experiences of women in general, rather than solely those of the mentally ill.[46] However, the association of gender with the experiences and representations of mental illness remains a key area of interest, with an ongoing expansion of research into the connections between mental disorders and masculinity.[47]

[45] Tomes, *A Generous Confidence,* pp. 28, 190, ch. 3 and 5; Joan Busfield, "The female malady? Men, women and madness in nineteenth-century Britain", *Sociology*, 28 (1994), pp. 259–77; Aude Fauvel, "Madness: A 'female malady'? Women and psychiatric institutionalisation in France", in Patrice Bourdelais and John Chircop (eds), *Vulnerabilities, Social Inequalities and Health in Perspective*, Évora: Ediçoes Colibri, 2010, pp. 61–75; Anatole Le Bras, "Troubles dans la masculinité", *Histoire, Médecine et Santé*, 23 (2023), pp. 55–69.

[46] Janice Chesters, "A horror of the asylum or of the home: Women's stories 1880–1910", in Catherine Coleborne and Dolly MacKinnon (eds), *Madness in Australia: Histories, Heritage and the Asylum*, St. Lucia: University of Queensland Press, 2003, pp. 135–44; Jill Julius Matthews, *Good and Mad Women: The Historical Construction of Femininity in Twentieth-Century Australia,* Sydney: George Allen & Unwin, 1984; Wallis, "Hysterical women".

[47] See, for example, Hewitt, *Institutionalizing Gender*; John Starrett Hughes, "The madness of separate spheres: Insanity and masculinity in Victorian Alabama", in Mark Carnes and Clyde Griffen (eds), *Meanings for Manhood: Constructions of Masculinity in Victorian America*, Chicago: University of Chicago Press, 1990, pp. 53–66; Mark Micale, *Hysterical Men. The Hidden History of Male Nervous Illness*, Cambridge, MA: Harvard University Press, 2008; Christopher Forth, "Neurasthenia and manhood in fin-de-siècle France", in Gijswijt-Hofstra and Porter (eds), *Cultures of Neurasthenia*, pp. 329–61; Kostas Makras, "'The poison that upsets my reason': Men, madness and drunkenness in the Victorian period", in Thomas Knowles and Serena Trowbridge, *Insanity and the Lunatic Asylum in the Nineteenth Century*, London: Pickering & Chatto, 2015, pp. 135–48; Helen Goodman, "'Madness and masculinity'. Male patients in London asylums and Victorian culture", in Trowbridge and Knowles (eds), *Insanity and the Lunatic Asylum in the Nineteenth Century*, pp. 149–66; Camille Bajeux and Aude Fauvel, "'The nervous testicle'. Urology and psychiatry in Belle Époque France", *Histoire, Médecine et Santé*, 23 (2023), pp. 35–54. As Le Bras convincingly argues, the psychiatric hospital played a normalising role for both female and male behaviour. Le Bras, "Troubles dans la masculinité". Similar to history, the sociology of mental health and illness has expanded its interest in men and masculinities in recent years. See, indicatively, Margaret Alston and Jenny

In comparison with the extensive international historiography on mental illness and psychiatry, studies on Greece are relatively scarce. The majority of these studies have been produced by psychiatrists and they do not always take into account the theoretical and methodological developments in the historiography of psychiatry. They concentrate on the history of psychiatric concepts, the progressive establishment of psychiatry and its relationship to medicine,[48] and the history of specific institutions—sometimes as experienced by the authors themselves.[49] While the social environment is occasionally mentioned, particularly in relation to the philanthropic activities of wealthy donors or the traditional response of Greek society to psychiatry, important topics like popular perceptions of madness, the role and place of psychiatry and psychiatric institutions within society, and the social, cultural and gendered meanings of mental illness have not been subjected to systematic investigation, with few

Kent, "The big dry: The link between rural masculinities and poor health outcomes for farming men", *Journal of Sociology*, 44 (2), 2008, pp. 133–47; Ester Carolina Apesoa-Varano, Judith C. Barker and Ladson Hinton, "'If you were like me, you would consider it too': Suicide, older men, and masculinity", *Society and Mental Health*, 8 (2), 2018, pp. 157–73; Jonathan Scourfield, "Suicidal masculinities", *Sociological Research Online*, 10 (2), 2005.

[48] Thanasis Karavatos, "Η γένεση της ψυχιατρικής και του ασύλου. Οι μυθοποιήσεις της πραγματικότητας και ο πραγματικός μύθος" ["The genesis of psychiatry and the asylum. The mythologisations of reality and the real myth"], *Ο Πολίτης [The Citizen]*, 78 (1987), 52–55; Thanasis Karavatos, *Τα Πρώτα Βήματα της Ψυχολογίας στην Ελλάδα. Η «Ψυχολογία» του Γεωργίου Σερουΐου [The First Steps of Psychology in Greece. The "Psychology" of Yeorgios Serouios], Athens: Οδυσσέας [Odysseas],* 1991; Grigoris Ampatzoglou, *Ψυχιατρική και Ιατρική. Τα Όρια, οι Ταυτότητες και οι Σχέσεις [Psychiatry and Medicine. The Boundaries, the Identities and the Relationships]*, Athens: Οδυσσέας [Odysseas], 1991; Dimitris Ploumpidis, *Ιστορία της Ψυχιατρικής στην Ελλάδα. Θεσμοί, Ιδρύματα και Κοινωνικό Πλαίσιο 1850-1920 [History of Psychiatry in Greece: Institutions, Hospitals and Social Framework 1850–1920]*, Athens: Εξάντας [Exandas], 1995. The latter has been the main secondary source for most historical studies of Greek psychiatry, including the present one, but also for sociological and anthropological works, including Amy Blue, *Η Δημιουργία της Ελληνικής Ψυχιατρικής [The Creation of Greek Psychiatry]*, Athens: Εξάντας [Exandas], 1999; Manolis Tzanakis, "Μετασχηματισμοί υποκειμενικότητας και καθημερινότητα. Μια κοινωνιολογική προσέγγιση της ψυχιατρικής μεταρρύθμισης στην Ελλάδα" ["Transformations of subjectivity and everyday life. A sociological approach of psychiatric reform in Greece"], PhD thesis, University of Crete, 2003.

[49] Nikolaos Tsikis (ed.), *Δρομοκαΐτειο Θεραπευτήριο, 1887–1987 + 15. Εκατό Δεκαπέντε Χρόνια Κοινωνικής Προσφοράς [Dromokaitio Sanatorium, 1887–1987 + 15. One Hundred Fifteen Years of Social Offer]*, Athens: s.n., 2001; Kostas Filandrianos, *Δημόσιο Ψυχιατρείο Αθηνών. Το Δαφνί, μια Φανταστική Πολιτεία [The Public Psychiatric Hospital of Athens. Dafni, an Imaginary City]*, Athens: s.n., 1977.

works adopting a social history perspective.[50] Nevertheless, over the past twenty years there has been a notable increase in interest in the history of the "psy" sciences and mental health. Significant studies have been conducted on the history of psychoanalysis, psychiatric reform and specific institutions, notably the psychiatric hospital of Leros.[51] However, the historiography of mental health in Greece remains incomplete. This book attempts to address these shortcomings by engaging with the contemporary international historiography on the relationship between psychiatric institutions and their social environment.

[50] Notable examples are the works of Vangelis Karamanolakis and Panayota Kazolea-Tavoulari. Vangelis Karamanolakis, "Η ανάδυση των ιδρυματικών θεσμών. Το Δρομοκαΐτειο φρενοκομείο (1880–1903)" ["The emergence of institutions. The Dromokaitio Lunatic Asylum (1880–1903)"], MA thesis, Athens, National and Kapodistrian University of Athens, 1997; Vangelis Karamanolakis, "Το Δρομοκαΐτειο φρενοκομείο: 1887–1903. Όψεις της εγκατάστασης ενός ιδρυματικού θεσμού" ["The Dromokaitio Lunatic Asylum: 1887–1903. Aspects of the establishment of an institution"], *Μνήμων [Mninon]*, 20 (1998), pp. 45–66; Vangelis Karamanolakis, "Οι φρενοβλαβείς εν ταις οδοίς" ["The lunatics in the streets"], in *Η Πόλη στους Νεότερους Χρόνους: Μεσογειακές και Βαλκανικές Όψεις (19ος-20ός Αιώνας) [The City in Modern Times: Mediterranean and Balkan Aspects (19th-20th Century)]*, Proceedings of the 2nd International Conference of History, Athens: Εταιρεία Μελέτης Νέου Ελληνισμού [Society for the Study of Modern Hellenism], 2000, pp. 193–202; Panayota Kazolea-Tavoulari, "Άσυλο φρενοβλαβών Σύρου: Ιστορική διερεύνηση προνοιακών θεσμών και κοινωνικές αναπαραστάσεις περί ψυχικής νόσου" ["Syros Lunatic Asylum: Historical investigation of welfare institutions and social representations of mental illness"], *Τα Ιστορικά [Ta Historica]*, 19 (2002), pp. 377–98.

[51] See, for example, Lena Atzina, *Η Μακρά Εισαγωγή της Ψυχανάλυσης στην Ελλάδα [The Long Introduction of Psychoanalysis in Greece]*, Athens: Εξάντας [Exandas], 2004; Danae Karydaki (ed.), *Η Λέρος στο Επίκεντρο και το Περιθώριο [Leros in the Spotlight and the Margin]*, Athens: Ψηφίδες [Psiphides], 2020; Despo Kritsotaki, *"Ψυχική Υγιεινή", Κοινωνική Πρόνοια και Ψυχιατρική Μεταρρύθμιση στη Μεταπολεμική Ελλάδα. Το Κέντρο Ψυχικής Υγιεινής και Ερευνών, 1956–1978 ["Mental Hygiene", Social Welfare and Psychiatric Reform in Post-War Greece. The Centre for Mental Health and Research, 1956–1978]*, Athens: Πεδίο [Pedio], 2016; Despo Kritsotaki, "Σεξουαλικότητα και επιστήμες 'ψ' στην Ελλάδα: Έφηβες και νέες γυναίκες σε μια υπηρεσία ψυχικής υγείας, 1960–1980" ["Sexuality and 'psy' sciences in Greece: Adolescent girls and young women in a mental health service, 1960–1980"], in Dimitra Vassiliadou and Glafki Gkotsi (eds), *Ιστορίες για τη Σεξουαλικότητα [Histories about Sexuality]*, Athens: Θεμέλιο [Themelio], 2020, pp. 81–100; Despo Kritsotaki, "Changing psychiatry or changing society? The "Motion for the Rights of the 'Mentally Ill' in Greece, 1980–1990", *Journal for the History of Medicine and Allied Sciences*, 2021.

Sources and Methodology

In order to achieve this objective, my research draws on hitherto unexplored material from the archives of two psychiatric institutions, the Eginitio and the Dromokaitio. As is the case with psychiatric institutions throughout the world, both institutions have generated and preserved extensive archives over the course of their history. In this study, I have used two main types of archival documents: the admission or patient registers (hereafter referred to as "registers") and the patient history case notes (hereafter "case histories").

The registers were organised by date of admission and recorded a range of data pertaining to the patients. This included basic demographic information such as name, age and occupation, as well a medical data such as diagnosis and outcome of hospitalisation. Given the considerable number of patients admitted to the institutions during the period under study (1900–1920), the analysis was conducted on a sample comprising all patients admitted during the years 1905–1909 and 1916–1920 (Table 1.1).[52] The samples are representative of the total patient admissions, as they are large and have been drawn in a systematic manner. Additionally, the trends indicated are consistent with those reported for the entire hospital population in other sources, such as annual reports.

The samples given in Table 1.1 include patients who were registered multiple times because they were readmitted to the same institution. The repetition of the same patients' data provides a complete picture of admissions but disturbs the picture of the patients themselves. For example, a female patient who was admitted ten times to the Eginitio between the years 1905 and 1909 would be perceived as ten distinct women of a similar age and with the same diagnosis, marital status and occupation. For this reason, since the study focuses on patients rather than admissions, patients who were recorded in the registers more than once, that is to say those who were readmitted, were removed from the samples.[53]

[52] The choice of two sub-periods was based on the availability of registers in the Eginitio at the outset of the research project. At that time, only the registers for the years 1904–1909 and 1916–1930 were available, while the register for the intermediate years (1910–1915) was missing. This register was eventually located, but the processes of data collection and database construction had already been completed.

[53] Given the prevalence of the phenomenon of readmissions, the same data processing of the records of psychiatric institutions has been conducted by other researchers as well. For example, Anne Digby, "Quantitative and qualitative perspectives on the asylum", in

Table 1.1 Number of patients recorded in the registers of the Eginitio and the Dromokaitio

Years	*Patients*	
	Eginitio	*Dromokaitio*
1905–1909	985	465
1916–1920	571	611
Complete period	1556	1076

Sources: *Α΄ Μητρώον Εσωτερικών Ασθενών Αιγινητείου Νοσοκομείου* [*First Register of Inpatients of the Eginitio Hospital]*; *Γ΄ Μητρώον Εσωτερικών Ασθενών Αιγινητείου Νοσοκομείου* [*Third Register of Inpatients of the Eginitio Hospital*], archive of the Eginitio Hospital, Eg/1 and Eg/1.1, Athens. *Βιβλίον των κατ' Αίτησιν Ιδιωτών Εισαγομένων Φρενοβλαβών [Book of the Insane Admitted at the Request of Private Individuals]*, 1904–1906, 1906–1909, 1909–1911, 1915–1917, 1917–1918, 1918–1919, 1919–1921; *Α΄ Βιβλίον των εξ Επαγγέλματος Εισαγομένων Ψυχοπαθών* [*First Book of ex Officio Admitted Psychopaths]* and *Β΄ Βιβλίον των εξ Επαγγέλματος Εισαγομένων Ψυχοπαθών [Second Book of ex Officio Admitted Psychopaths]*, archive of the Dromokaitio Hospital, Athens.

The samples obtained through this methodology are the focus of analysis in this book (Table 1.2).

The second principal source of analysis in this study is case histories. This type of document has been employed with increasing frequency in historical research over the past three decades to investigate the experiences of patients, the meanings and treatments of mental illness, and

Table 1.2 Patients' samples with and without readmissions

Institution	*Patients' number*	
	With readmissions (original sample)	*Without readmissions (final sample)*
Eginitio	1556	1430
Dromokaitio	1076	961

Sources: See Table 1.1.

Roy Porter and Andrew Wear (eds), *Problems and Methods in the History of Medicine*, London: Croom Helm, 1987, pp. 153–74.

the role of gender and family in mental health care.[54] Furthermore, case histories have been examined as scientific texts that document the genesis and evolution of psychiatric theory and practice, as well as the invention and representation of the self in modernity.[55] Finally, patient histories are particularly useful in exploring everyday clinical practices, which do not necessarily align with medical theory as presented in scholarly publications, including monographs, articles and textbooks.[56] Despite the varying methodologies employed by historians in analysing case notes, the historiography has established that by the second half of the nineteenth century, the practices of recording psychiatric histories had become systematised, with different institutions in different countries maintaining their patients' histories in a comparable manner.

This was also the case in Greece. In the Eginitio and the Dromokaitio, case histories were recorded in registers until the first decades of the twentieth century. The number of pages per case varied according to the amount of information available, the length of the patient's stay and the interest of the medical staff. In the Eginitio, for the period 1900–1920, only the case histories from the years 1907–1909 are preserved, which I studied in their entirety (263 case histories). In the Dromokaitio, where the set of case histories (more than 1000) is intact, I had to limit my

[54] Jonathan Andrews, "Case notes, case histories and the patients' experience of insanity at Gartnavel Royal Asylum, Glasgow, in the nineteenth century", *Social History of Medicine*, 2 (1998), pp. 255–81; Houston, *Madness and Society in Eighteenth-Century Scotland*, pp. 173–236; Catharine Coleborne, *Reading "Madness": Gender and Difference in the Colonial Asylum in Victoria, Australia, 1848–1888*, Perth: API Network, 2007; Tanja Luckins, "'Crazed with grief?' The asylum and the Great War in Australia", in Coleborne and MacKinnon (eds), *Madness in Australia*, pp. 169–80; Matthews, *Good and Mad Women*; Hilary Marland, *Dangerous Motherhood: Insanity and Childbirth in Victorian Britain*, Houndmills: Palgrave Macmillan, 2004, pp. 97–105; Marjorie Levine-Clark, "Dysfunctional domesticity: Female insanity and family relationships among the West Riding poor in the mid-nineteenth century", *Journal of Family History*, 25 (2000), pp. 341–61; Alexandra Wallis, "Driven to insanity: Marital cruelty and the female patients at the Fremantle Lunatic Asylum, 1858–1908", *Limina*, 24 (2019), pp. 1–16.

[55] John Forrester, "If p, then what? Thinking in cases", *History of the Human Sciences*, 9 (1996), pp. 1–25; Matt Reed, "Making the case: The evolution of the case study in nineteenth-century French psychiatry", Ph.D. thesis, Faculty of Claremont Graduate University, 2002. The latter however is based on published case histories and not on case notes.

[56] Günter Risse and John Harley Warner, "Reconstructing clinical activities; patients records in medical history", *Social History of Medicine*, 2 (1992), pp. 183–205.

research to a systematic sample (195 case histories), which gives a representative picture of the whole and facilitates the analysis, as these are often long texts with a wealth of information.

The first part of the case histories describes each patient's condition prior to admission: heredity, previous illnesses, and information on character, habits and lifestyle. The time of onset, symptoms, causes and course of the illness that brought the patient to the hospital were then recorded, as well as the patient's condition on admission ("present condition"). The last part of the case histories deals with the patient's condition during hospitalisation. This section is not particularly well developed and provides scarce and usually standardised information. On the contrary, the first and second sections of the case histories offer fascinating insights into the social roles of psychiatric hospitals and psychiatry and lend themselves to the analysis of the main themes of this book: under what circumstances and for what reasons was someone considered to be in need of psychiatric hospitalisation, and by whom and why was admission to the institutions decided.

Importantly, the case histories of the Eginitio and the Dromokaitio support the contention of researchers working with case histories, that one of the key strengths of these sources is that they were produced through the psychiatrists' encounters with patients and their private and public environments. As a result, case histories can reveal the multiplicity of relationships and forces that shaped the operation of the institutions, as well as the ways in which the institutions were treated by different subjects, including patients. It becomes clear that case histories should be analysed as polyphonic sources, reflecting not only medical views but, to a certain extent, lay narratives: those of the patients, their relatives or others who met and spoke with the medical staff of the hospitals. This makes case histories one of the few windows into lay perceptions of mental illness and psychiatric institutions.[57]

[57] Jonathan Sadowsky, *Imperial Bedlam: Institutions of Madness in Colonial Southwest Nigeria*, Berkeley: University of California Press, 1999, p. 49; Sally Wilde, "The elephants in the doctor-patient relationship: Patients' clinical interactions and the changing surgical landscape of the 1890s", *Health and History*, 9 (2007), pp. 2–27; Michael MacDonald, "Madness, suicide, and the computer", in Roy Porter and Andrew Wear (eds), *Problems and Methods in the History of Medicine*, pp. 207–29; Catharine Coleborne, "'His brain was wrong, his mind astray': Families and the language of insanity in New South Wales, Queensland, and New Zealand, 1880s–1910", *Journal of Family History*, 31 (2006), pp. 45–65.

However, the different voices, medical and non-medical, were not equally represented in the case histories. The finished story was inevitably composed by the medical staff of the institutions, who admitted that they treated lay testimonies with suspicion,[58] often took the patients' statements as evidence of illness,[59] guided the informants and decided what to record and how.[60] Through language and selection, medical interpretation and ideology were built into case histories.[61] Nevertheless, we cannot ignore the lay narratives within the case histories, even if they are only "fragmented statements" entangled with the medical ones.[62] Ultimately, the case histories were the product of a dynamic interaction between the lay and professional worlds, which obliges us to study both sides together.[63] After all, the distinction between medical and lay views of mental illness should not be seen as absolute, as there were continuity and exchange between them. On the one hand, the doctors' interests and ideology were constituted through their contact with the patients and their milieu and within a specific social and cultural environment

[58] For example, Thomas Clouston, *Clinical Lectures on Mental Diseases*, 5th edition, London: J. Churchill, 1898, pp. 123–24; John Charles Bucknill and Daniel Hack Tuke, *A Manual of Psychological Medicine: Containing the History, Nosology, Description, Statistics, Diagnosis, Pathology, and Treatment of Insanity: with an Appendix of Cases*, London: J. Churchill, 1858, pp. 271–72; Maurice Craig, *Psychological Medicine. A Manual of Mental Diseases for Practitioners and Students*, 3rd edition, London: J. Churchill, 1917, p. 415.

[59] Martine Delvaux, *Femmes Psychiatrisées, Femmes Rebelles: De l'Étude de Cas à la Narration Autobiographique*, Paris: Le Plessis-Robinson, Institut Synthelabo, 1998, pp. 65–67; Benoît Majerus, *Parmi les Fous: Une Histoire Sociale de la Psychiatrie au XXe Siècle*, Rennes: Presses Universitaires de Rennes, 2019, p. 24.

[60] Grigoris Ampatzoglou, "Αυτοβιογραφία, μνήμη, θεραπεία" ["Autobiography, memory, therapy"], in Rika Benveniste and Theodoros Paradellis (eds), *Διαδρομές και Τόποι της Μνήμης: Ιστορικές και Ανθρωπολογικές Προσεγγίσεις [Routes and Places of Memory: Historical and Anthropological Approaches]*, Athens: Αλεξάνδρεια [Alexandria], 1999, pp. 65–87.

[61] See Andrews, "Case notes, case histories and the patients' experience", pp. 255–281, which also highlights that since the 1870s the direct testimony of patients disappears from case histories due to the increasing emphasis on organic causes and physical symptoms of illness.

[62] MacDonald, "Madness, suicide, and the computer", p. 211.

[63] Marie-Claude Thifault and Isabelle Perreault, "Behind asylum walls: Studying the dialectic between psychiatrists and patients at Montreal's Saint-Jean-de-Dieu Hospital during the first half of the twentieth century", *Social History of Medicine*, 32 (2019), pp. 377–94.

that encompassed the medical and scientific fields.[64] On the other hand, professional views were not only influenced by social and cultural experiences but also shaped them. In fact, especially from the end of the nineteenth century, medical theories of mental illness were widespread in broad segments of society.[65]

Of course, as with any historical source, we need to determine the reliability of case histories. Were inaccurate or untrue problems deliberately noted in them to justify the incarceration of people who were not mentally ill? This question is not asked as part of the wider issue—not addressed in this book—of whether or not patients in psychiatric institutions were actually mentally ill. True or not, the problems outlined in the case histories were presented as sufficient reasons for psychiatric hospitalisation and thus reflected medical and lay perceptions of what problems could or should be treated in psychiatric hospitals. However, the possibility of invented mental health problems is instructive in terms of the use of institutions: if confirmed, it shows that psychiatric institutions were not (only) used to treat mental illness but (also) for other reasons and purposes. Although no evidence was found in this research that the disorders recorded in the case histories were untrue, it was revealed that they were related not only to mental but also to physical problems, as well as to behaviours that were difficult, uncontrollable and incompatible with the norms of the time (see parts two and three of the book). Therefore, regardless of whether the patients' mental symptoms were real or not, it seems that the definitions of mental illness were quite broad or that the use of institutions was not limited to the treatment of mental illness.

In addition to the registers and case histories, my research includes other sources from the institutions. For the Dromokaitio, I have taken into account letters sent to the institution by patients' relatives, which I found in the case histories, as well as annual reports. The latter provided the public with information on the operation and finances of the institution and were aimed at attracting private patients and sponsors for the

64 Coleborne, "His brain was wrong, his mind astray"; Wilde, "The elephants in the doctor-patient relationship"; David Wright, "'Childlike in his innocence'. Lay attitudes to 'idiots' and 'imbeciles' in Victorian England", in David Wright and Anne Digby (eds), *From Idiocy to Mental Deficiency. Historical Perspectives on People with Learning Disabilities*, London: Routledge, 1996, pp. 118–33.

65 Lynn Gamwell and Nancy Tomes, *Madness in America: Cultural and Medical Perceptions of Mental Illness before 1914*, Binghamton: Cornell University Press, 1995, pp. 87–89; Reed, "Making the case", p. 1.

hospital. As such, they tend to present an idealised picture of the institution, but also contain useful information about the patient population and the aims and ideology of the institution's medical and administrative authorities. Finally, for the Eginitio, which was a university clinic, I also analysed administrative and financial documents from the archives of the Rectorate of the University of Athens (the *Clinics-Hospitals* series).

Apart from the archives of the institutions, I use psychiatric textbooks of the period, both Greek and foreign that were internationally recognised at the beginning of the twentieth century. While these texts represent the most conservative elements of scientific research and late guides to shifts in scientific paradigms,[66] they can provide insights into the established psychiatric theory on which doctors were trained and which served as a basis for their practice. The textbooks contain not only the established categories of illness and their clinical descriptions, but also rich material on the social (epidemiological) characteristics of the mentally ill, which can be analysed to reveal how physicians approached and understood, among other things, the issues of gender, age and social class in relation to mental illness.

Terminology

Although it has been argued that studies of the history of psychiatry cannot speak of mental illness in general because this overlooks the marked differences between different illnesses,[67] the general term mental illness is helpful because it encompasses a number of medical and lay terms that were in use in the early twentieth century, and which I also use in this book, such as lunacy, madness and insanity. Without considering what these conditions would be called today, I employ the generic term mental illness to denote the set of problems for which treatment in psychiatric institutions or lunatic asylums, as they were still called in Greece, was considered appropriate in the early decades of the twentieth century. At the same time, I also examine separately the specific diagnostic categories used by specialist physicians (see Chapter 8).

[66] Vatz and Weinberg, "The rhetorical paradigm in psychiatric history", pp. 311–30, referring to the paradigm shift theory by Thomas Kuhn.

[67] Trevor Turner, "A diagnostic analysis of the casebooks of Ticehurst House Asylum, 1845–1890", *Psychological Medicine Supplements*, 21 (1992), pp. 1–70.

Another issue of terminology concerns the term used to describe the branch of medicine that specialised on mental illness. Although what we now call psychiatry had acquired the status of an independent medical specialty in many countries by the 1880s,[68] and the terms "psychiatrist" and "psychiatry" can be found in the international literature as early as the first half of the nineteenth century, these terms were rarely used by either physicians or the public until the end of the nineteenth century.[69] In Greece, at that time, the term "psychiatrist" gradually replaced the older term "phrenologist".[70] The shift in terminology was evidenced by the university nomenclature: in 1891 the chair of "Neurology and Phrenic Diseases" was established in the Medical School of the University of Athens, but its holder, Michael Katsaras, was "Professor of Neurology and Psychiatry", while his book, the first textbook on the subject in Greek, published between 1898 and 1900, was entitled *Pathology of the Nerves and Psychiatry*. The period under consideration is therefore a transitional one in terms of how the mental health expertise and experts were referred to. The modern terms "psychiatry" and "psychiatrist", which are used throughout this book mainly for reasons of convenience, were not fully established, but were introduced and were increasingly used in the early twentieth century.

Finally, the definition of psychiatry in relation to neurology needs to be clarified. From the end of the eighteenth century, psychiatry was in the process of detaching itself from general medicine, mainly by having its own professional space, the asylum. Neurology, on the contrary, continued to be practised chiefly in the context of pathology: neurologists and "nerve doctors", although a distinct category from the eighteenth

[68] Albrecht Hirschmüller and Magda Whitrow, "The development of psychiatry and neurology in the nineteenth century", *History of Psychiatry*, 10 (1999), pp. 395–423; William Bynum, "The nervous patient in 18th- and 19th-century Britain: The psychiatric origins of British neurology", in Robin Murray and Trevor Turner (eds), *Lectures on the History of Psychiatry: The Squibb Series*, London: Gaskell, 1990, pp. 115–27.

[69] The German-speaking countries were an exception, as the terms "Psychiatrie" and "Psychiater" were used already from the middle of the nineteenth century. Aude Fauvel, "Psychiatrie", *Bulletin de la Société Française pour l' Histoire des Sciences de l' Homme*, 29 (2006), pp. 43–51.

[70] The term "phrenologist" comes from the word "phren" ("φρην", plural "φρένες"), meaning the mind, brain and reason. From the same word come the terms "φρενοπάθεια" and "φρενοβλάβεια" (illness and damage of the mind), which were commonly used in the nineteenth and in the early twentieth centuries to refer to mental illness.

century, remained primarily pathologists working in private practice or in general hospitals. However, in the second half of the nineteenth century, psychiatry and neurology began to overlap to some extent in several countries. In German-speaking countries, the two disciplines were practised in the same institutions—universities, clinics and laboratories—and sometimes by the same people, the "neuropsychiatrists". In France, the most famous neurologist of the nineteenth century, Jean-Martin Charcot (1825–1893), who argued that neurology should be a separate medical specialty and identified himself as a neuropathologist, also treated psychiatric patients in his private practice and was posthumously considered a psychiatrist.[71] In Greece, the combination of psychiatry and neurology was evident in their academic organisation. As already noted, at the University of Athens at the beginning of the twentieth century, neurology and psychiatry shared the same chair, clinic and textbook, while Katsaras, the first holder of the chair, director of the clinic and author of the textbook, was a "neurologist-psychiatrist" who had trained in Paris in both fields.[72]

The academic and professional overlap between the two disciplines in Greece and internationally reflected the fluidity of their fields of knowledge, as some diseases were considered simultaneously or alternatively as psychiatric and neurological. This was the case for conditions with a known, even partially, neuropathology, but with prominent mental symptoms, such as general paralysis of the insane, and, conversely, for diseases without an identified organic basis but with marked physical ailments, such as hysteria, neurasthenia and hypochondria (see Chapter 8). These three diagnoses were also used, at least until the First World War, to describe disorders that lay between mental health and illness, and were usually treated by general practitioners, "nerve doctors" and neurologists in private practice, hydrotherapy establishments or neurological clinics.[73] Treatment outside asylums and the nerve idiom for conditions

[71] Toby Gelfand, "Neurologist or psychiatrist? The public and private domains of Jean-Martin Charcot", *Journal of the History of the Behavioral Sciences*, 36 (2000), pp. 215–29.

[72] Neurological Clinic, University of Athens, *Αιγινήτειον Νοσοκομείον 1905-2005 [Eginitio Hospital 1905–2005]*, Athens: s.n., 2005, p. 16. While separate chairs (and clinics) for psychiatry and neurology were founded in 1963, neurology and psychiatry remained a unified medical specialty until 1981.

[73] Edward Shorter, *A Historical Dictionary of Psychiatry*, Oxford: Oxford University Press, 2005, p. 22–26, 129; German Berrios, *The History of Mental Symptoms: Descriptive*

with mental symptoms were justified by the psychiatric theory of the time that all mental disorders were caused by brain or nerve damage, even if unknown.[74] At the same time, the widespread revulsion and fear of madness and of the asylum, and the stigma associated with them, perpetuated the tendency towards neurological vocabularies and cures. Unlike mental illness, nervous illness suggested physical ailments, which made it more reassuring and acceptable. For the same reason, some physicians chose to identify themselves as nerve specialists, neurologists, neuropsychologists or neuropsychiatrists, even if treating mental patients, in order to attract clients and gain higher social status.[75]

The evolution of neurology and its relationship with psychiatry, a sometimes cooperative and sometimes competitive relationship, is a long and complex subject that cannot be analysed here.[76] Nonetheless, it can be stated with a reasonable degree of certainty that at the beginning of the twentieth century, psychiatry and neurology were not entirely distinct disciplines, but rather, they had some degree of overlap. Therefore, during this period, neurology and what was termed "nervous diseases" cannot be excluded from the study of psychiatry and mental illness.

Psychopathology since the Nineteenth Century, Cambridge: Cambridge University Press, 1995, p. 316; Janet Oppenheim, *"Shattered Nerves": Doctors, Patients, and Depression in Victorian England*, New York: Oxford University Press, 1991, p. 5.

74 This theory had been supported since the late eighteenth century but was strengthened from the mid-nineteenth century. For instance, Wilhelm Griesinger, one of the best-known proponents of the organicity of mental illness, conceived mental illnesses as brain diseases, a theory that equated mental patients with nervous patients and unified psychiatry and neurology. Shorter, *A History of Psychiatry*, pp. 27–28; Edward Shorter, *From Paralysis to Fatigue: A History of Psychosomatic Illness in the Modern Era*, New York: Free Press, 1992, p. 208; Paul Lerner, *Hysterical Men: War, Psychiatry, and the Politics of Trauma in Germany, 1890–1930*, Ithaca, N.Y.: Cornell University Press, 2003, p. 84. For an in-depth examination of the localisation theory, see Katja Guenther, *Localization and Its Discontents: A Genealogy of Psychoanalysis and the Neuro Disciplines,* Chicago: University of Chicago Press, 2015.

75 Shorter, *A History of Psychiatry*, pp. 113–16, 136–44.

76 Lucy Ozarin, "The American Neurological Association: Early days and relations with psychiatry", *Psychiatric News*, 42 (2007), p. 11; Bonnie Ellen Blustein, "'A hollow square of psychological science': American neurologists and psychiatrists in conflict", in Scull (ed.), *Madhouses, Mad-doctors, and Madmen*, pp. 241–70; Shorter, *A History of Psychiatry*, pp. 142–43.

Synopsis

The book is divided into three parts, corresponding to three levels of analysis. The first part presents the social and institutional background of the study, examining the social conditions in Greece at the beginning of the twentieth century and the main features of mental health care at that time. In particular, Chapter 2 deals with those aspects of Greek social history that are most closely linked to the way mental illness and psychiatric institutions were perceived: the growth of the urban population; the characteristics of social welfare and charities; the roles of family; the social position of women and children. Chapter 3 turns to the official framework of mental health care, namely psychiatric legislation and psychiatric institutions. Focusing on the two hospitals studied, it shows that the Eginitio and the Dromokaitio were very different from each other, but their functioning was determined by a set of common actors. These actors, as well as the profile of each institution, are further outlined in Chapter 4, which analyses the economic parameters of psychiatric hospitalisation—how much it cost and who paid for it.

In the second part, the analysis moves to the level of the mentally ill. It examines how early twentieth-century psychiatric theory conceptualised the social and medical characteristics of the mental patients and investigates the relevant information recorded in the Eginitio and the Dromokaitio. Focusing on the categories of gender (Chapter 5), age and marital status (Chapter 6), occupation class and place of residence (Chapter 7), diagnosis and symptoms (Chapter 8), the analysis sheds light on medical and lay perceptions of mental illness, the social and cultural roots and implications of medical discourses, and the ways in which the institutions were used and operated. Who was admitted to mental hospitals and why? Were the institutions socially acceptable, that is, were they incorporated into public and private strategies for dealing with people who were considered mentally ill, difficult or different, and on what terms? By addressing these questions, Chapters 5, 6, 7 and 8 each contribute in their own way to an understanding of the social functions of psychiatry and psychiatric institutions.

This is the focus of the third part, which is devoted to the level of communication between the institution and society. Chapters 9 and 10 look at the involvement of patients and their private and public environment in hospital care, inquiring who made the decision to admit patients and why, who was responsible for them, who brought them to the hospital and provided information about them, and who requested their discharge. Chapter 11 considers four other indicators of the relationship between the institutions and society: the duration of illness before admission, the length of stay in hospital, the methods of treatment before and after admission, and visits to psychiatric hospitals by private and public actors, such as patients' relatives, representatives of public authorities and journalists. The final chapter 12 maps out another function of the institutions, tracing their role in achieving the professional and scientific objectives of psychiatry.

Overall, the three parts of the book interact to raise the issue of the social position of the mentally ill and psychiatric institutions. Taken together, the three levels of analysis (social and institutional context, psychiatric patients and the interaction between institutions and society) reveal the dynamic processes that shaped psychiatric hospitalisation: the negotiations, compromises and ruptures between private, public and medical actors in accordance with the changing and multifaceted needs of each. Ultimately, the book illuminates the complex ways in which mental illness was perceived and treated in early twentieth-century Greece, the continuities and discontinuities, tensions and interactions between experts and non-experts, including the patients and their milieu in both the private and the public sphere.

PART I

Mental Health Care in Context

To understand the multiple experiences of mental illness and psychiatric hospitalisation, we need to place them in their social and institutional context. To this end, Chapter 2 examines those aspects of early twentieth-century Greek society that are most directly related to the social meanings of mental illness and psychiatric hospitalisation, including urbanisation, welfare policies, and gender and family roles. Chapter 3 traces the development of psychiatric institutions in Greece, with reference to psychiatric legislation and the establishment of the first mental hospitals. It outlines the origins and problems of mental health care and the differences between the two institutions studied, the Eginitio and the Dromokaitio. These differences are further highlighted in Chapter 4, which focuses on the economic aspects of institutional mental health care. The cost of hospitalisation and the agents, private and public, who paid the hospital fees, shaped both the terms of access to the institutions and the profile of the Eginitio and the Dromokaitio, and introduce the multiple agents involved in psychiatric hospitalisation, an issue that will be discussed in more detail in the third part of the book.

CHAPTER 2

Greece in the Early Twentieth Century

The first two decades of the twentieth century were an turbulent period for Greece, marked by successive wars (the Balkan Wars, the First World War, the Greco-Turkish War), population movements and the National Schism (the violent conflict between the monarchists and the Venizelists, named after the leader of the Liberal Party, Eleftherios Venizelos).[1] Rather than dwelling on the numerous events of the period 1900–1920, the following brief account focuses on aspects of Greek social history that relate to the themes of the book, namely the profile of the population, the process of urbanisation, the family, and gender roles.

At the beginning of the century, Greece was undergoing radical change. Between 1907 and 1920, with the annexation of Crete, Epirus, Macedonia, Thrace and the islands of the eastern Aegean, the country's territory and population more than doubled (from 63,211 to 149,150 square kilometres and from 2,631,952 to 5,531,474 inhabitants, of whom 2,865,463 lived in "Old Greece"). At the same time, cities and their population grew. While the country remained largely rural—only

[1] For a brief, informative account of the history of this period in English, see Richard Clogg, *A Concise History of Greece*, 4th edition, Cambridge: Cambridge University Press, 2021, ch. 3. On Venizelos, see Paschalis Kitromilidis, *Eleftherios Venizelos: The Trials of Statesmanship*, Edinburgh: Edinburgh University Press, 2006.

D. Kritsotaki, *A Social History of Mental Health Care in Greece, c. 1900–1920*, Mental Health in Historical Perspective,
https://doi.org/10.1007/978-3-032-05764-8_2

about one in four Greeks lived in cities of 5,000 inhabitants or more—the rural population and employment in agriculture, husbandry and fishing had declined compared with the nineteenth century. The increase in the urban population was most marked in the industrial cities, such as Piraeus, Athens, Volos, Chalkida and Kalamata, and was due to the rural exodus that had begun at the end of the previous century, the settlement of large numbers of refugees from Crete, Eastern Romelia, Macedonia, Asia Minor and the Caucasus, and the incorporation of the New Territories, especially Macedonia, which included twenty-two cities with a population of over 5,000 inhabitants. The largest Greek cities in 1907 were Athens (167,479 inhabitants), Piraeus (71,505 inhabitants) and Patras (37,724 inhabitants), to which Thessaloniki was added in 1920, after the Balkan Wars with more than 200,000 inhabitants. In that year, there were 1.3 million people living in cities of 5,000 inhabitants or more, about half of them in the three largest cities: Athens, Thessaloniki and Piraeus.[2]

The capital, Athens, expanded irregularly in all directions. In 1907, it was five times larger than in 1879, covering 1,918 hectares.[3] Its population had grown from 44,510 inhabitants in 1870 to 292,991 in 1920.[4] Athens was the political, administrative and economic centre of Greece.

[2] *Στατιστικά Αποτελέσματα της Γενικής Απογραφής του Πληθυσμού κατά την 27 Οκτωβρίου 1907 [Statistical Results of the General Census of the Population on 27 October 1907]*, v. 1, Athens: Εθνικό Τυπογραφείο [National Press], 1909, p. ξδ; *Απογραφή του Πληθυσμού της Ελλάδος κατά την 19 Δεκεμβρίου 1920. Γενικά Στατιστικά Αποτελέσματα [Census of the Population of Greece on 19 December 1920. General Statistical Results]*, Athens: Εθνικό Τυπογραφείο [National Press], 1928, p. ο; Christos Hadziiossif, "Η μπελ επόκ του κεφαλαίου" ["The belle époque of capital"], in Christos Hadziiossif (ed.), *Ιστορία της Ελλάδας του 20ού Αιώνα [History of Greece in the 20th Century]*, v. A1, Athens: Βιβλιόραμα [Vivliorama], 1998, pp. 309–49; Socrates Petmezas, "Η δημογραφική συγκυρία: Η δεύτερη φάση της 'δημογραφικής μετάβασης' και η υπερατλαντική μετανάστευση" ["The demographic conjuncture: The second phase of the 'demographic transition' and transatlantic migration"], in Hadziiossif (ed.), *History of Greece in the 20th Century*, v. A1, pp. 41–51; Christina Agriantoni, "Βιομηχανία" ["Industry"], in Hadziiossif (ed.), *History of Greece in the 20th Century*, v. A1, pp. 173–221; Christina Agriantoni, "Η ελληνική οικονομία. Η συγκρότηση του ελληνικού καπιταλισμού, 1870–1909" ["The Greek economy. The constitution of Greek capitalism, 1870–1909"], in Vasilis Panayotopoulos (ed.), *Ιστορία του Νέου Ελληνισμού, 1770–2000 [History of Modern Hellenism, 1770–2000]*, v. 5, Athens: Ελληνικά Γράμματα [Ellinika Grammata], 2003, pp. 55–70.

[3] Aleka Karadimou-Gerolympou, "Πόλεις και πολεοδομία" ["Cities and urban planning"], in Hadziiossif (ed.), *History of Greece in the 20th Century*, v. A1, pp. 223–53.

[4] Christos Loukos, "Η ελληνική πρωτεύουσα, 1890-1912" ["The Greek capital, 1890–1912"], in Christina Koulouri (ed.), *Αθήνα, Πόλη των Ολυμπιακών Αγώνων 1896-1906*

Although its main supplier was Piraeus, the hub of Greek industry and commerce, Athens had an extensive industrial zone and concentrated the country's main financial institutions—the stock exchange, banks and company headquarters. From the end of the nineteenth century, Athens gradually transformed itself into a European capital thanks to infrastructure works and new buildings such as administrative, educational and athletic facilities, hotels, museums, theatres and urban mansions.[5]

Overall, by the beginning of the twentieth century, the urban sector had developed and was the driving force of the economy. The number of workers in industry and commerce had increased[6] and a large—if we include the petty bourgeoisie—and politically, economically and socially active middle class had emerged in the urban centres. Owners of small or large professional units, producers and/or salesmen, public

[Athens, City of the Olympic Games 1896-1906], Athens: Διεθνής Ολυμπιακή Ακαδημία [International Olympic Academy], 2004, pp. 55–98.

[5] Agriantoni, "The Greek economy"; Christina Agriantoni, *Οι Απαρχές της Εκβιομηχάνισης στην Ελλάδα τον 19ο Αιώνα [The Beginnings of Industrialisation in Greece in the 19th Century]*, Athens: Εμπορική Τράπεζα της Ελλάδος [Commercial Bank of Greece], 1986, pp. 99–104, 339–42; Yannis Yannitsiotis, *Η Κοινωνική Ιστορία του Πειραιά, 1860–1910. Η Συγκρότηση της Αστικής Τάξης [The Social History of Piraeus, 1860–1910. The Formation of the Urban Class]*, Athens: Νεφέλη [Nefeli], 2006; Lila Leontidou, *Πόλεις της Σιωπής. Εργατικός Εποικισμός της Αθήνας και του Πειραιά, 1909–1940 [Cities of Silence. Labour Settlement of Athens and Piraeus, 1909-1940]*, Athens: Πολιτιστικό Ίδρυμα Ομίλου Πειραιώς [Piraeus Bank Group Cultural Foundation], 2013; Loukos, "The Greek Capital, 1890–1912"; Maro Kardamitsi-Adami, "Ο αστικός χώρος, 1871–1909. Πολεοδομία-αρχιτεκτονική" ["The urban space, 1871–1909. Urban planning-architecture"], in Panayotopoulos (ed.), *History of Modern Hellenism, 1770–2000*, v. 5, pp. 107–26; Vasiliki Theodorou, "Αλητεία και επαιτεία στην Αθήνα και τον Πειραιά. Οι πρώτες απόπειρες 'αναμόρφωσης παραστρατημένων παίδων' (τέλη 19ου αρχές 20ού αιώνα)" ["Truancy and begging in Athens and Piraeus. The first attempts to 'reform misguided children' (late 19th and early twentieth century)"], in *Η Πόλη στους Νεότερους Χρόνους: Μεσογειακές και Βαλκανικές Όψεις (19ος-20ός Αιώνας) [The City in Modern Times: Mediterranean and Balkan Aspects (19th-20th Century)]*, Proceedings of the 2nd International Conference of History, Athens: Εταιρεία Μελέτης Νέου Ελληνισμού [Society for the Study of Modern Hellenism], 2000, pp. 203–30.

[6] Leda Papastefanaki, "Μισθωτή εργασία" ["Paid work"], in Kostas Kostis and Socrates Petmezas (eds), *Η Ανάπτυξη της Ελληνικής Οικονομίας κατά τον 19ο αιώνα (1830–1914) [The Development of the Greek Economy in the nineteenth century (1830–1914)]*, Athens: Αλεξάνδρεια [Alexandria], 2006, pp. 253–91; Agriantoni, "Industry"; Agriantoni, "The Greek economy".

and private servants, military personnel, estate agents, teachers, engineers, doctors, bank managers and lawyers were distinguished themselves from the popular classes on the basis of property, education or lifestyle, adopting "European" habits and emphasising the values of work, economy, discipline and progress.[7]

Furthermore, by the early twentieth century there was a large labour force massed in the cities. Although still to a certain extent temporary and mobile, and often dependent on the rural economy and traditional kinship and local labour relations, this labour force was certainly more consolidated than in the previous century and had begun to form a working class. It was a period of radicalisation for both the urban and rural workers, and of the first peasant and labour movements and trade unionism—intensified by the miserable living and working conditions, exhausting hours and inadequate incomes.[8] The problems were serious in both rural and urban areas, but urban poverty caused more concern, being perceived as a social threat directly linked to begging, prostitution, vagrancy and crime.[9] Beyond the moral connotations of poverty in discourses of the

[7] Yannitsiotis, *The Social History of Piraeus, 1860–1910,* pp. 78–80, 110–11, 185; Loukos, "The Greek capital, 1890–1912"; Mark Mazower, "The messiah and the bourgeoisie: Venizelos and politics in Greece, 1909–1912", *The Historical Journal*, 35 (1992), pp. 885–904.

[8] Petros Pizanias, *Οι Φτωχοί των Πόλεων: Η Τεχνογνωσία της Επιβίωσης στην Ελλάδα το Μεσοπόλεμο [The Urban Poor: The Expertise of Survival in Greece in the Interwar Period]*, Athens: Θεμέλιο [Themelio], 1993, pp. 13–19, 99–100, 110; Antonis Liakos, *Εργασία και Πολιτική στην Ελλάδα του Μεσοπολέμου: Το Διεθνές Γραφείο Εργασίας και η Ανάδυση των Κοινωνικών Θεσμών [Labour and Politics in interwar Greece: The International Labour Office and the Emergence of Social Institutions]*, Athens: Εμπορική Τράπεζα της Ελλάδος [Commercial Bank of Greece], 1993, pp. 85–94; Agriantoni, "The Greek economy"; Yannitsiotis, *The Social History of Piraeus, 1860–1910*, pp. 91, 284–85, 299; Socrates Petmezas, "Αγροτική οικονομία" ["Rural economy"], in Hadziiossif (ed.), *History of Greece in the 20th Century*, v. A1, pp. 53–85; Keti Tsichli-Aroni, "Το αγροτικό ζήτημα. Θεσσαλία- Νέες Χώρες, 1909–1922" ["The agricultural issue. Thessaly—New Territories, 1909–1922"], in Vasilis Panayotopoulos (ed.), *Ιστορία του Νέου Ελληνισμού, 1770–2000 [History of Modern Hellenism, 1770–2000]*, v. 6, Athens: Ελληνικά Γράμματα [Ellinika Grammata], 2003, pp. 267–82; Agriantoni, "Industry"; Gelina Harlaftis, "Ο εμπορικός στόλος στους πέντε ωκεανούς και τις πέντε ηπείρους" ["The merchant fleet in the five oceans and five continents"], in Vasilis Panayotopoulos (ed.), *Ιστορία του Νέου Ελληνισμού, 1770–2000 [History of Modern Hellenism, 1770–2000]*, v. 6, Athens: Ελληνικά Γράμματα [Ellinika Grammata], 2003, pp. 129–38.

[9] Maria Korasidou, *Οι Άθλιοι των Αθηνών και οι Θεραπευτές τους: Φτώχεια και Φιλανθρωπία στην Ελληνική Πρωτεύουσα τον 19ο Αιώνα [The Miserable of Athens and their Healers: Poverty and Charity in the Greek Capital in the 19th Century]*, Athens:

middle and upper classes, it is undeniable that the urban lower classes faced intense problems, which the bourgeoisie witnessed daily, because, unlike in the major cities of other countries, they continued to live in the centre of Athens and Piraeus—few districts were purely working class.[10] In the early twentieth century, the constant wars and the lack of infrastructure in the cities to accommodate the refugees and immigrants swelled the problems: unemployment, inflation, lack of housing, high rents, overcrowding, poor sanitary conditions, inadequate nutrition, endemic and pandemic diseases, high mortality rates, increased numbers of orphans and unprotected children.[11]

Attempts to deal with urban social problems had already been made in the first half of the nineteenth century, when begging and vagrancy were criminalised and the Ministry of the Interior began to distribute financial aid to the poor, albeit in an unstable and unsystematic way. At the same time, public health legislation was introduced and hospitals for the poor were established. In the second half of the nineteenth century, much of the welfare work was carried out by philanthropists and charitable associations, who, sometimes in cooperation with the municipal authorities, organised soup kitchens and established institutions, such as the Athens poorhouse, schools and workshops for destitute women, workers and children, and hospitals, including the Dromokaitio Lunatic Asylum.[12]

Κέντρο Νεοελληνικών Ερευνών, Εθνικό Ίδρυμα Ερευνών [Centre for Modern Greek Studies, National Hellenic Research Foundation], 1995, pp. 71–97; Yannitsiotis, *The Social History of Piraeus, 1860–1910,* pp. 241–59; Yeorgios Kokkinos, "Η αμφιθυμία για την πόλη. Η ελληνική διανόηση στην καμπή του 19ου αιώνα και το αστικό φαινόμενο" ["Ambivalence about the city. Greek intellectuals at the turn of the nineteenth century and the urban phenomenon"], in *The City in Modern Times*, pp. 597–614.

[10] Yannitsiotis, *The Social History of Piraeus, 1860–1910,* pp. 225–27.

[11] Kardamitsi-Adami, "The urban space"; Loukos, "The Greek capital, 1890–1912"; Karadimou-Gerolympou, "Cities and urban planning"; Liakos, *Labour and Politics*, pp. 34, 80–82; Yannitsiotis, *The Social History of Piraeus, 1860–1910*, pp. 269–74. From 1906 to 1920, about 800,000 refugees arrived in Greece, but a significant number of them emigrated to the USA and France. Nikos Andriotis, "Η πρώτη προσφυγιά. Ελληνικές προσφυγικές μετακινήσεις, 1906–1922" ["The first refugees. Greek refugee movements, 1906–1922"], in Panayotopoulos (ed.), *History of Modern Hellenism, 1770–2000]*, v. 6, pp. 95–104.

[12] Korasidou, *The Miserable of Athens*, pp. 28–31, 40–56, 72–82; Vasiliki Theodorou, ""Φιλανθρωπία και πόλη. Ορφανοί και άστεγοι παίδες στον Πειραιά γύρω στο 1875" ["Charity and the city. Orphans and homeless children in Piraeus around 1875"], *Μνήμων [Mnimon]*, 14 (1992), pp. 71–90.

In the first decades of the twentieth century, the state responded to the socio-economic problems aggravated by the wars and the arrival of refugees by organising national welfare institutions through the Ministry of Welfare, founded in 1917, but still largely involving and working with the private sector.[13] Meanwhile, in response to the escalating labour issue, and mainly to prevent social conflict and the radicalisation of workers, labour legislation was introduced in the 1910s to regulate working hours and conditions, child and female labour, and issues of safety, health and trade unionism. Although limited and poorly implemented, the new state welfare and labour policies signalled the expansion of state intervention, a trend observed in many European countries in the early twentieth century.[14]

State intervention also targeted women. The new legislation in the 1910s that regulated women's work was motivated by concerns about prostitution, venereal disease and the moral and physical condition of the nation. These were grounded on accepted notions of gender attributes and roles, which in the nineteenth and early twentieth century were determined by the ideal of domesticity and separate spheres not only in the middle but also in the lower classes. The latter were infiltrated by bourgeois conceptions of gender through literature and the press, the education system, bourgeois philanthropy and the legislation on women's and children's work itself. The ideology of domesticity and separate spheres conceptualised the private sphere of the home as a refuge, a place of harmonious and altruistic relationships, which was identified with women. It was represented as strictly separated from the public—and masculine, competitive and immoral—sphere of politics and the market. In addition, the ideology of the family wage, based on men's work,

[13] Theodorou, "Truancy and begging"; Michael Ailianos, *Το Έργον της Ελληνικής Περιθάλψεως [The Work of Greek Welfare]*, Athens: s.n., 1921, pp. 68–75, 169–70; Vasiliki Theodorou, "Μεταβαλλόμενα πλαίσια συνάφειας μεταξύ εθελοντών, ειδικών και κράτους: Το παράδειγμα του Πατριωτικού Ιδρύματος Προστασίας του Παιδιού" ["Changing contexts of affinity between volunteers, experts and the state: The example of the Patriotic Foundation for the Protection of Children"], in Efi Avdela, Haris Exertzoglou, Christos Lyrintzis (eds), *Μορφές Δημόσιας Κοινωνικότητας στην Ελλάδα του Εικοστού Αιώνα [Forms of Public Sociality in Twentieth Century Greece]*, Rethymno: Πανεπιστήμιο Κρήτης [University of Crete], 2015, pp. 82–100.

[14] Christos Hadziiossif, "Εισαγωγή" ["Introduction"], in Hadziiossif (ed.), *History of Greece in the 20th Century*, v. A1, pp. 9–39.

justified the devaluation of women's work, which was defined as complementary to men's, while long and hard hours of work were perceived as physically and morally harmful to women and their maternal duties, as well as to children.[15]

Despite the strength of the ideology of domesticity and separate spheres, the roles of women were multiple and complex.[16] In Greece, as in many other countries, women were heads of families, managers of family businesses, owners and workers.[17] According to the censuses, which usually underestimate the extent of women's work, the female

[15] Aleka Boutzovi, "Γυναικείο κίνημα. Όψεις και δράσεις, 1909–1922" ["Women's movement. Aspects and actions, 1909–1922"], in Panayotopoulos (ed.), *History of Modern Hellenism, 1770–2000*, v. 6, pp. 283–92; Eleni Varika, *Η Εξέγερση των Κυριών. Η Γένεση μιας Φεμινιστικής Συνείδησης στην Ελλάδα, 1933–1907 [The Rising of the Ladies. The Genesis of a Feminist Consciousness in Greece, 1933–1907]*, Athens: Κατάρτι [Katarti], 2004, pp. 66–70, 106–7, 148–52, 183; Angelika Psarra, "Μητέρα ή πολίτις: Ελληνικές εκδοχές της γυναικείας χειραφέτησης (1870–1920)" ["Mother or citizen: Greek versions of female emancipation (1870–1920)"], *Δίνη [Dini]* 9 (1997), pp. 239–57; Eleni Fournaraki, "Περί μορφώσεως 'χρηστών μητέρων' και εκπαιδεύσεως 'μελλόντων πολιτών': Έμφυλοι λόγοι στην ελληνική εκπαίδευση τον 19ο αιώνα" ["On the formation of 'righteous mothers' and the education of 'future citizens': Gendered discourses in Greek education in the nineteenth century"], in Vasiliki Theodorou and Vasiliki Kontoyianni (eds), *Το Παιδί στη Νεοελληνική Κοινωνία, 19ος-20ός αιώνας. Αξίες, Αναπαραστάσεις, Αποτυπώσεις [The Child in Modern Greek Society, Nineteenth-Twentieth century. Values, Representations, Impressions]*, Athens: Ελληνικό Λογοτεχνικό και Ιστορικό Αρχείο [Hellenic Literary and Historical Archive], 1999, pp. 73–120. The seminal work on the ideology of separate spheres and domesticity in the nineteenth century is Leonore Davidoff and Catherine Hall, *Family Fortunes: Men and Women of the English Middle Class, 1780–1850*, London: Hutchinson, 1987.

[16] See, for example, Jo Burr Margadant, "Introduction: Constructing selves in historical perspective", in Jo Burr Margadant (ed.), *The New Biography: Performing Femininity in Nineteenth-Century France*, Berkeley: University of California Press, 2000, 1–32; Andrea Mansker, *Sex, Honor and Citizenship in Third Republic France*, London: Palgrave Macmillan, 2011.

[17] Yannitsiotis, *The Social History of Piraeus, 1860–1910*, pp. 183–201; Evridiki Sifnaiou, *Έλληνες Έμποροι στην Αζοφική. Η Δύναμη και τα Όρια της Οικογενειακής Επιχείρησης [Greek Merchants in the Sea of Azov. The Power and Limits of Family Enterprise]*, Athens: Ινστιτούτο Νεοελληνικών Ερευνών/ Εθνικό Ίδρυμα Ερευνών [Institute of Modern Hellenic Studies/ National Hellenic Research Foundation], 2009, ch. 7; Zizi Salimpa, *Γυναίκες Εργάτριες στην Ελληνική Βιομηχανία και στη Βιοτεχνία (1870–1922) [Women Workers in Greek Industry and Manufacture (1870–1922)]*, Athens: Ινστιτούτο Νεοελληνικών Ερευνών/ Εθνικό Ίδρυμα Ερευνών [Institute of Modern Hellenic Studies/ National Hellenic Research Foundation], 2002; Maria Preka, "Αρσακειάδες δασκάλες (1880–1913): Ρητές και άρρητες οικογενειακές διαδρομές" ["Teachers of the Arsakeio School (1880–1913): Explicit and implicit family paths"], in Glafki Gotsi, Androniki Dialeti and Eleni

working population had risen from 4.4% of the total female population in 1907 to 9.6% in 1920.[18] Women were mainly employed as servants and industrial workers. (Lower) middle-class women found a principal occupational outlet in teaching, and from the early twentieth century gradually entered the public sector, for example in telegraph offices. Beyond paid employment, women were active in the public sphere through charitable work and participation in voluntary associations. Especially in the interwar period, a growing number of women's associations could be described as feminist: they aimed to improve women's rights to education and work, albeit without questioning the perceived natural destiny of women as housewives and mothers. Some of these associations also promoted women's political rights, which, however, were not won until after the Second World War.[19] In addition, in the first decades of the twentieth century women had more opportunities to move out of the

Fournaraki (eds), *Το Φύλο στην Ιστορία. Αποτιμήσεις και Παραδείγματα [Gender in History. Assessments and Examples]*, Athens: Ασίνη [Asini], 2015, pp. 107–38.

[18] *Statistical Results of the General Census of the Population on 27 October 1907*; *Census of the Population of Greece on 19 December 1920*. Historians have convincingly argued that official censuses in Greece and other countries in the nineteenth and early twentieth centuries underrated women's work, mainly because they did not record atypical and unregistered occupations, which accounted for a large part of women's work. Leda Papastefanaki, "Ανάμεσα στην ιστορία της βιομηχανίας και την ιστορία της εργασίας. Η οπτική του φύλου στην ελληνική ιστοριογραφία" ["Between the history of industry and the history of labour. The gender perspective in Greek historiography"], in Gotsi, Dialeti and Fournaraki (eds), *Gender in History*, pp. 79–106.

[19] Papastefanaki, "Paid work"; Maria Repousi, "Το φύλο των γυναικών. Έμφυλες ομάδες και γυναικείες διεκδικήσεις" ["The gender of women. Gendered groups and women's demands", in Panayotopoulos (ed.), *History of Modern Hellenism, 1770–2000*, v. 5, pp. 187–96; Yannitsiotis, *The Social History of Piraeus, 1860–1910*, pp. 59 and 286–87; Boutzovi, "Women's movement. Aspects and actions, 1909–1922"; Efi Avdela, *Δημόσιοι Υπάλληλοι Γένους Θηλυκού. Καταμερισμός της Εργασίας κατά Φύλα στον Δημόσιο Τομέα, 1908–1955 [Female Public Servants: Labour Division by Gender in the Public Sector, 1908–1955]*, Athens: Ίδρυμα Έρευνας και Παιδείας της Εμπορικής Τράπεζας της Ελλάδος [Research and Education Foundation of the Commercial Bank of Greece], 1990; Efi Avdela, "Οι γυναίκες, κοινωνικό ζήτημα" ["Women, a social issue"], in Christos Hadziiossif (ed.), *Ιστορία της Ελλάδας του 20ού αιώνα 1922–1940, Ο Μεσοπόλεμος [History of Greece in the 20th Century 1922–1940, The Interwar Period]*, v. B1, Athens: Βιβλιόραμα [Vivliorama], 2002, pp. 337–59; Pothiti Hantzaroula, *Σμιλεύοντας την Υποταγή. Οι Έμμισθες Οικιακές Εργάτριες στην Ελλάδα το Πρώτο Μισό του Εικοστού Αιώνα [Forging Submission. Paid Domestic Workers in Greece in the First Half of the Twentieth Century]*, Athens: Παπαζήσης [Papazisis], 2012.

private sphere of the family in their social interactions and recreational practices.[20]

Still, in the early twentieth century women's dependence and incapacity were taken for granted as natural and were enshrined in law. Women were largely identified with the family, even if ideologically, an identification reinforced by the low percentage of declared female employment and unmarried women, as well as the tendency for women to stop working systematically outside the home after marriage. This trend was linked to the size of families, which on average had more children than in Western Europe, but also to the considerable social functions of the family.[21] The latter was the main means of coping with poverty, illness and old age, while public welfare was limited and mainly covered those who had no family or could not be supported by it. The family network of support extended beyond the nuclear family, although care for the extended family seems to have declined among the urban population from the late nineteenth century onwards.[22] In any case, the main issue was less about the self-sufficiency of the individual than that of the family, which, at least in the lower social classes, relied on the income and contributions of all its members.[23]

Relationships and roles in the family are a multifaceted issue, with many more aspects than suggested by the almost non-existent divorce rate, the loving and selfless relationships projected by the domestic ideal, and the inequality grounded on the legal, social and cultural supremacy of men and the dependence of women and children.[24] The complexity of family

[20] Dimitra Vassiliadou, *Στον Τροπικό της Γραφής. Οικογενειακοί Δεσμοί και Συναισθήματα στην Αστική Ελλάδα, 1850–1930 [In the Tropic of Writing. Family Ties and Emotions in Urban Greece, 1850–1930]*, Athens: Gutenberg, 2018, pp. 142–160.

[21] Petmezas, "The demographic conjuncture"; Salimpa, *Women Workers in Greek Industry and Manufacture*, p. 88.

[22] Yannitsiotis, *The Social History of Piraeus, 1860–1910,* pp. 197–98.

[23] Pizanias, *The Urban Poor*, pp. 133; Kostas Fountanopoulos, "Η μισθωτή εργασία" ["Paid work"], in Hadziiossif (ed.), *History of Greece in the 20th Century*, v. A1, pp. 87–121; Liakos, *Labour and Politics*, pp. 77–79.

[24] In 1920, only 0.2% of the population over 14 were divorced. *Census of the Population of Greece on 19 December 1920,* p. LXXXII. On the multiple meanings of family relationships in late nineteenth- and early twentieth-century Greece, see Vassiliadou, *In the Tropic of Writing*. For an international perspective, Michelle Perrot, "Roles and characters", in Philippe Aries and Georges Duby (eds), *A History of Private Life*, v. 4, Cambridge, MA.: Harvard University Press, 1987, pp. 167–259 and Gunilla-Friederike Budde, *Auf dem*

relations is illustrated by the example of parents-children relationships: the heightened concern for children in the context of the domestic ideal meant more care for them and, at the same time, greater attempts to control them.[25] In addition to the family, other actors sought to shape the lives and the behaviour of children in the early twentieth century: the state, with legislation on child labour and general restrictions on children's rights; the press, which stressed the social dangers faced by street children and criticised their inappropriate behaviour; and the particularly authoritarian school system of the time.[26] However, child labour was widespread and only a minority of children attended school, even though primary education had been compulsory since 1834.[27] At the beginning of the twentieth century, the literacy rate was quite low—33.7% in 1907 and 40.1% in 1920—and even lower among women, whose access to primary and especially secondary and higher education was limited.[28]

In summary, Greece in the first decades of the twentieth century was a rural country with urban and industrial enclaves. Nonetheless, it experienced problems and concerns similar to those of more urbanised and industrialised countries, albeit to different extent and intensity. These were addressed by private philanthropy rather than state welfare. The basic institution for care and support remained the family, in which gender and age roles were largely regulated by the ideology of separate spheres and domesticity, but in practice were far more complex. Against this background, the following chapters will examine the interaction of psychiatric discourses and institutions with welfare practices, family functions,

Weg ins Bürgerleben: Kindheit und Erziehung in Deutschen und Englischen Bürgerfamilien, 1840–1914, Gottingen: Vandenhoeck & Ruprecht, 1994.

[25] Yannitsiotis, *The Social History of Piraeus, 1860–1910,* pp. 193 and 376.

[26] Fountanopoulos, "Paid work"; Yannitsiotis, *The Social History of Piraeus, 1860–1910,* pp. 152, 193–96, 230, 376; Theodorou, "Truancy and begging".

[27] Yannis Lixouriotis, *Κοινωνικές και Νομικές Αντιλήψεις για το Παιδί τον Πρώτο Αιώνα του Νεοελληνικού Κράτους [Social and Legal Perceptions of the Child in the First Century of the Modern Greek State]*, Athens/Ioannina: Δωδώνη [Dodoni], 1986, pp. 64–110.

[28] *Statistical Results of the General Census of the Population on 27 October 1907,* p. ξ; *Census of the Population of Greece on 19 December 1920*, p. 11. In the late nineteenth and early twentieth centuries, women made up a small proportion of university students (1.4%), but their numbers were growing. Sidiroula Ziogou-Karastergiou, *Γυναίκες και Ανώτατη Εκπαίδευση στην Ελλάδα. Οι Πρώτες Φοιτήτριες στο Πανεπιστήμιο Αθηνών (1890–1920) [Women and Higher Education in Greece. The First Female Students at the University of Athens (1890–1920)]*, Thessaloniki: n.p., 1988, p. 50.

and gender, age and class roles in early twentieth-century Greece. First, however, it is necessary to outline the context of mental health care in which the two institutions under study were situated.

CHAPTER 3

Mental Health Care in Greece up to the First Decades of the Twentieth Century

The Eginitio and the Dromokaitio operated within the legal and institutional framework for mental illness and the mentally ill that began to take shape in the second half of the nineteenth century, when psychiatric legislation was introduced, and the first psychiatric institutions were established in Greece. These developments were linked to the psychiatric reform movement that had taken place in several European countries a few decades earlier: laws were enacted to create a system of psychiatric institutions and to regulate the legal, administrative and bureaucratic issues arising from state intervention in matters of mental illness.[1] Three factors played an important role in this European trend: the mobilisation of psychiatrists who argued that insanity could only be cured in a new type of institution—the asylum; concerns about personal rights and the need for protection from unjust psychiatric confinement; and, finally, as will be illustrated in the case of Greece, by state efforts to control public expenditure on the "insane".[2]

[1] For example, on France, see Robert Castel, *The Regulation of Madness: The Origins of Incarceration in France*, Cambridge: Polity, 1988 and on England, Roy Porter, *Mind Forg't Manacles: A History of Madness in England from the Restoration to the Regency*, London: Athlone, 1987, pp. 1–32.

[2] Edward Shorter, *A History of Psychiatry: From the Era of the Asylum to the Age of Prozac*, New York: John Wiley, 1997, pp. 8–17; Maurice Craig, *Psychological Medicine.*

D. Kritsotaki, *A Social History of Mental Health Care in Greece, c. 1900–1920*, Mental Health in Historical Perspective,
https://doi.org/10.1007/978-3-032-05764-8_3

Yet, while psychiatric reform was underway, earlier approaches, such as seeing mental illness as caused by evil spirits, and practices, such as exorcisms, home isolation and physical restraint of the mentally ill, continued to be part of how mental illness was being understood and managed in the nineteenth and twentieth centuries.[3] Especially in rural Greece, religious practices, including pilgrimages and confinement in monasteries, were still popular in the early twentieth century.[4] However, although lay and religious concepts and practices existed in parallel or in conflict with medical ones, the latter grew stronger: the care of the mentally ill gradually shifted to psychiatric institutions, which, as we shall see below, expanded in the early twentieth century.

At the time, law 742 "On the Establishment of Lunatic Asylums" of 1862 was still in force. It was the result of the discussions about the issue of the mentally ill and the first attempts to establish a psychiatric hospital in the late 1850s and early 1860s.[5] An inquiry conducted by the Ministry of the Interior in 1858 sheds light on the official political and medical discourse on the issue of the "insane" in the Greek state in mid-nineteenth century. It reported that 422 "lunatics" were living with their families, wandering freely, living in monasteries, being held in police stations, or being treated in psychiatric clinics abroad. All these practices were denounced as inadequate or scandalous, with the exception of treatment

A Manual of Mental Diseases for Practitioners and Students, 3rd edition, London: J. Churchill, 1917, p. 444; Joshua John Schwieso, "Religious fanaticism and wrongful confinement in Victorian England", *Social History of Medicine,* 9 (1996), pp. 159–74.

[3] John Walton, "Casting out and bringing back in Victorian England: Pauper lunatics, 1840–70", in William Frederick Bynum, Roy Porter and Michael Shepherd (eds), *The Anatomy of Madness: Essays in the History of Psychiatry,* v. 2, London: Tavistock, 1985, pp. 132–46; Peter Bartlett and David Wright, "Community care and its antecedents", in Peter Bartlett and David Wright (eds), *Outside the Walls of the Asylum: On 'Care and Community' in Modern Britain and Ireland*, New Brunswick: Athlone Press, 1999, pp. 1–18; Robert Allan Houston, *Madness and Society in Eighteenth-Century Scotland*, Oxford: Clarendon Press, 2000, pp. 391–92; Muriel Bonaïti, "Histoire de l'asile XIXe siècle: D'un lieu unique de soin aux premières alternatives à l'hospitalisation", Ph.D. thesis, Université de Lyon, 2007, p. 91.

[4] Dimitris Ploumpidis, *Ιστορία της Ψυχιατρικής στην Ελλάδα. Θεσμοί, Ιδρύματα και Κοινωνικό Πλαίσιο 1850- 1920 [History of Psychiatry in Greece: Institutions, Hospitals and Social Framework 1850–1920]*, Athens: Εξάντας [Exandas], 1995, pp. 23–39. As we shall see, some of the patients admitted to the Eginitio had spent some time in monasteries before being hospitalised.

[5] Ibid., pp. 142–66.

abroad, which was only available to a few wealthy Greeks.[6] The report accompanying the inquiry adopted a humanist rhetoric—emphasising the deplorable condition of the mentally ill—to justify the need for a mental health law and a public mental hospital. At the same time, it is possible to discern three motives behind the mobilisation of the state: firstly, the need to take measures to deal with the "dangerous insane"; secondly, the high cost—subsidised by the state—of maintaining the mentally ill in monasteries; thirdly, the effort to modernise and "Europeanise" Greece.[7] Although a public mental hospital was not founded until the beginning of the twentieth century, the law passed in 1862, modelled on the French one, constituted a statement that Greece was looking towards European policies for the treatment of the mentally ill.

As the name of the law suggests, its first provision was to allow the government to establish state "lunatic asylums" as needed. The asylums were placed under the "supreme direction of the Government" and more specifically under the Ministry of the Interior, but the public authorities that would be responsible for founding them were not specified. The law regulated the administration and funding of mental hospitals, and identified the public authorities that supervised them and made unannounced visits. It dealt with the issues of the patients' liberty and property and outlined the legal and medical procedures for admission and discharge. Two types of hospital placement were set out: by private application and by "order of the authority". Hospitals were required to notify the public authorities of all admissions and to keep separate registers for each category of patients. In the first type of placement, the patient's private environment, a relative, friend or other acquaintance, applied for admission, and the public authorities (usually the Mayor or the police) intervened only if the applicant was illiterate and unable to write and sign the application. In the second type of placement, the role of the authorities was more pronounced. This placement concerned patients "whose mental illness could pose a threat to public order or the personal safety

[6] Ibid., pp. 127, 136.

[7] Ibid., p. 128; Amy Blue, *Η Δημιουργία της Ελληνικής Ψυχιατρικής [The Creation of Greek Psychiatry]*, Athens: Εξάντας [Exandas], 1999, pp. 61–62; Manolis Tzanakis, "Μετασχηματισμοί υποκειμενικότητας και καθημερινότητα. Μια κοινωνιολογική προσέγγιση της ψυχιατρικής μεταρρύθμισης στην Ελλάδα" ["Transformations of subjectivity and everyday life. A sociological approach of psychiatric reform in Greece"], Ph.D. thesis, University of Crete, 2003, p. 67.

of citizens". "Ex officio patients", as they were called, were admitted in mental hospitals in Athens and Piraeus by order of the chief of police and in other regions by order of the Prefect or, in urgent cases, the Mayor.[8] Apart from public officials, medical experts were also involved in the placement process, as two medical certificates from doctors not affiliated to the mental hospital were required for admission.[9]

The coexistence of private, public and medical actors was also evident in the legislative arrangements for discharging patients. The law stipulated that all discharges were overseen by public authorities, but the decision to discharge could be taken by different actors. Patients were discharged as soon as the hospital doctor confirmed that they had been cured and entered this in the patients' register. Patients could also be discharged without being cured at the request of their guardian, a relative—first the spouse, then ascendants and direct descendants—or the person who had applied for admission; however, the doctor was entitled to refuse to discharge patients whom he considered dangerous and to keep them in the institution with the approval of the Prefect. Patients could be discharged as well by order of the Minister of the Interior or the Prefect. The latter had the main responsibility for discharging "ex officio patients", on the basis of the reports he received every three months from the doctor of the institution.[10]

Moving on from legislation to institutions, their establishment was a slow process. Law 742 did not to oblige the state to found "lunatic asylums", and in fact it was not until the mid-1910s that it did so.[11] The

[8] Νόμος ΨΜΒ' "Περί συστάσεως φρενοκομείων" [Law 742 "On the establishment of lunatic asylums", *Εφημερίς της Κυβερνήσεως [Government Gazette]*, 28, 19/5/1862. The Greek term is "ασθενείς εξ επαγγέλματος", which literally means "patients of profession" and is probably a translation of the French term "placements d' office". For this reason, I have chosen the translation "ex officio patients". This category of patients should not be confused with "public patients", a term I use to describe all patients whose fees were paid by public agents but who could have been admitted on private application.

[9] Ibid.

[10] Ibid.

[11] The discrepancy between the law and its implementation was not unique to Greece. In France, for example, the provision of the psychiatric law of 1838 for one institution per prefecture was not fulfilled, and many of the new institutions did not meet the requirements of the law (Georges Lanteri-Laura, "La chronicité dans la psychiatrie moderne française. Note d'histoire histoire théorique et sociale", *Annales. Économies, Sociétés, Civilisations*, 27 (1972), pp. 548–68). The difference is that in Greece the divergence between law and practice was much greater.

only public mental hospital until then was the Corfu psychiatric hospital, which had been founded in 1838 not by the Greek state but by the British during their rule in Corfu, and was transferred to Greece with the union of the Ionian Islands in 1864. By the beginning of the twentieth century, it had grown from a local to a national institution with more than 200 patients, mostly pauper, subsidised by the state, many of them labelled as "dangerous lunatics".[12] However, it was also opted for private placements, mainly because its fees were lower than those of the Dromokaitio and the Eginitio.[13] Private and public demand for inexpensive long-term hospitalisation meant that the institution took over the confinement of chronically ill patients, whereas earlier, in the 1870s, it had claimed a high scientific and social prestige.[14]

The Corfu hospital was not sufficient to meet the needs for low-cost and free psychiatric hospitalisation. People with mental health problems who were poor and without private support networks were often held in police stations after being arrested for vagrancy, disturbing public order or being considered dangerous. At the beginning of the twentieth century, in an attempt to change this situation, three makeshift asylums were founded in Athens, Souda (Crete) and Thessaloniki, which received mainly marginalised and "dangerous" patients, usually brought in by the police. The practice of detaining the mentally ill in police stations was weakened, but the conditions for the patients did not improve much. In Athens, on which we have more information, the "Insane Asylum", founded in Moschato in 1913 and moved to Kallithea at the end of the decade, was inadequate, filthy and police controlled; it offered no medical treatment and had no doctor until the 1920s. In 1925, the Legislative Decree "On the Organisation of the Public Psychiatric Hospitals of Athens, Thessaloniki and Chania" renamed the three asylums into

[12] Ploumpidis, *History of Psychiatry*, pp. 119–24, 166–82, 173–75; Anon., "Ειδήσεις: Το φρενοκομείο της Κερκύρας" ["News: The lunatic asylum of Corfu"], *Ψυχιατρική και Νευρολογική Επιθεώρησις [Psychiatric and Neurological Review]*, 6 (1903), p. 192.

[13] In the Corfu asylum, the minimum fee was 24 drachmas per month, in comparison with 60 in the Dromokaitio and the Eginitio. Ploumpidis, *History of Psychiatry,* p. 175 and here, ch. 4.

[14] Tzanakis, "Transformations of subjectivity and everyday life", pp. 57–58; Blue, *The Creation of Greek Psychiatry*, pp. 65–67.

"public psychiatric hospitals", established 400, 150 and 150 beds in each respectively and set the basis for their medical organisation.[15]

Beyond the Corfu psychiatric hospital and the three poorly organised public asylums, in early twentieth-century Greece there were a few small local asylums[16] and private clinics,[17] while mental patients were occasionally admitted to non-psychiatric institutions, such as the Asylum for the Incurable, and could also be examined and treated at the City Clinic in Athens.[18] But the main psychiatric institution since the late nineteenth century was the Dromokaitio. Founded in Athens in 1887 with a bequest from the businessman Zorzis Dromokaitis of Chios, it was later expanded

[15] Kostas Filandrianos, *Δημόσιο Ψυχιατρείο Αθηνών. Το Δαφνί, μια Φανταστική Πολιτεία [The Public Psychiatric Hospital of Athens. Dafni, an Imaginary City]*, Athens: s.n., 1977, pp. 21–47; Ilias Makris, *Οι Φρενοπαθείς Εγκληματίαι Αλλαχού και εν Ελλάδι [The Insane Criminals in Greece and Elsewhere]*, Athens: s.n., 1924, pp. 28–31.

[16] The local asylums included those of Kefallonia (founded by the British in 1840, with a capacity of 100 beds in 1938), Patras (founded in 1900), Syros (founded in 1908 on private and municipal initiative and accepting patients from all the Cycladic islands), Chania (founded in 1900), Chios (founded on private initiative in the 1880s) and the Jewish community of Thessaloniki (probably part of the Hirsch Hospital, founded in 1907). The latter three were transferred to the Greek state when the respective areas were incorporated into Greece after the Balkan Wars. Ploumpidis, *History of Psychiatry*, pp. 205–6, 208, 220; Maria Korasidou, *Όταν η Αρρώστια Απειλεί: Επιτήρηση και Έλεγχος της Υγείας του Πληθυσμού στην Ελλάδα του 19ου Αιώνα [When the Disease Threatens: Surveillance and Control of Population Health in 19th-Century Greece]*, Athens: Τυπωθήτω [Tipothito], 2002, pp. 98–99; Panayota Kazolea-Tavoulari, "Άσυλο φρενοβλαβών Σύρου: Ιστορική διερεύνηση προνοιακών θεσμών και κοινωνικές αναπαραστάσεις περί ψυχικής νόσου" ["Syros Lunatic Asylum: Historical investigation of welfare institutions and social representations of mental illness"], *Τα Ιστορικά [Ta Historica]*, 19 (2002), pp. 377–98.

[17] In Athens, in the 1910s, there were five private clinics. Ploumpidis, *History of Psychiatry*, pp. 211–19.

[18] The Asylum for the Incurable was a charitable institution founded in 1893 under the protection of the Queen (Anon, *Τα 80 Χρόνια του Ασύλου Ανιάτων: 1893–1973 [The 80 Years of the Asylum for the Incurable: 1893–1973*, Athens: s.n., s.d., pp. 13–15). Although its statute excluded the hospitalisation of the mentally ill (Asylum for the Incurable, *Άσυλον Ανιάτων, Καταστατικόν 1914* [*Asylum for the Incurable, Statute, 1914],* Athens: s.n., 1915, article 2), a small number of its inmates were affected by mental retardation and mental and nervous diseases (Asylum for the Incurable, *Άσυλο των Ανιάτων υπό την Προστασίαν της Α.Μ. της Βασιλίσσης, Έκθεσις των Πεπραγμένων του έτους 1907 [Asylum for the Incurable under the Protection of her Majesty the Queen, Report of the Activities of the year 1907]*, Athens: s.n., 1908, pp. 16–17). The institution is still in operation today. The City Clinic ("Αστυκλινική") was a general hospital that began operating in 1857 and offered free care to the poor when the Elpis Hospital, the municipal hospital of Athens, was full (Korasidou, *When the Disease Threatens*, pp. 83–86).

with other donations and bequests.[19] The Dromokaitio admitted private but also pauper patients—at the expense of the institution and its benefactors—[20]and ex officio patients, as it was subject to law 742. In addition, as we shall see in Chapter 7, it admitted patients from all over Greece.[21] In other words, although it was a private institution, the Dromokaitio had a public and national function. At the beginning of the twentieth century, 70 to 100 patients were admitted to the Dromokaitio each year and its population ranged from 300 to 330.[22]

The establishment of the Dromokaitio was presented as part of the wider effort to "modernise" and "Europeanise" the Greek state, society and mental health care.[23] The hospital was portrayed as a modern institution, where the latest scientific methods were applied by qualified and experienced staff, a model institution in late nineteenth- and early twentieth-century Greece. Indicatively, a 1914 brochure advertised the Dromokaitio as a therapeutic centre "perfected in accordance with the advances of Science", "one of the most perfect special therapeutic centres" in Greece with a hydrotherapy unit and a scientific laboratory.[24] This image is echoed in recent studies, which place the institution in the

[19] Ministry of the Interior, *Δημοσιεύματα του Τμήματος της Δημοσίας Οικονομίας [Publications of the Department of Public Finance]*, no. 4, Athens: s.n., 1899, pp. 288–89; Nikolaos Tsikis (ed.), *Δρομοκαΐτειο Θεραπευτήριο, 1887–1987 + 15. Εκατό Δεκαπέντε Χρόνια Κοινωνικής Προσφοράς [Dromokaitio Sanatorium, 1887–1987 + 15. One Hundred Fifteen Years of Social Offer]*, Athens: s.n., 2001, pp. 37–53.

[20] Νόμος ΑΦΙΔ' "Περί αδείας συστάσεως ιδιωτικού φρενοκομείου εκ του κληροδοτήματος Δρομοκαΐτου" [Law 1514 "On the licence for the establishment of a private lunatic asylum from the Dromokaitis bequest"], *Εφημερίς της Κυβερνήσεως [Government Gazette]*, 145, 10/6/1887.

[21] "Περί εγκρίσεως του προσωρινού Κανονισμού του συσταθέντος Φρενοκομείου του αοιδίμου Ζωρζή Δρομοκαΐτου" ["On the approval of the provisional Regulation of the established Lunatic Asylum of the Venerable Zorzis Dromokaitis"], *Εφημερίς της Κυβερνήσεως [Government Gazette]*, 192, 16/7/1887.

[22] Lucien Libert, "Les aliénés en Orient (Grèce, Empire Ottoman, Égypte)", *L'Informateur des Aliénistes et des Neurologistes*, Supplement de l' *Encéphale*, 7 (1912) pp. 283–302. The French doctor Libert travelled to Greece, the Ottoman Empire and Egypt in 1911–12, to observe how the mentally ill were treated in the "East", and published his impressions in the Supplement of the journal *Encéphale* in 1912–14.

[23] Tzanakis, "Transformations of subjectivity and everyday life", p. 59.

[24] Cited in Tsikis (ed.), *Dromokaitio Hospital, 1887–1987 + 15*, pp. 57–61. Libert, who visited the Dromokaitio in 1911, deemed it worthy of modern European institutions. Libert, "Les aliénés en Orient", p. 291.

scientific vanguard and emphasise that it was staffed by the "elite" of Greek psychiatrists who were active in the international scientific community.[25] However, we cannot readily accept the image of Dromokaitio as an exemplary institution, an image that was used by the management of the institution to attract private clients, whose fees were an important source of its income. Besides, this image did not reflect the full range of the institution's realities. At the beginning of the twentieth century, it was facing serious problems: the number of chronic and pauper patients was increasing, leading to overcrowding and financial difficulties.[26] Moreover, the opinion of the public about the institution was not entirely positive. According to the chairman of the board of directors in 1898, Yeorgios Makkas, the "people" treated the Dromokaitio like all psychiatric institutions: as a prison.[27]

In any case, the medical staff of the Dromokaitio included the most renowned Greek psychiatrists of the time: Christodoulos Tsirigotis (1841–1919), director from 1887 to 1903, who had studied in Italy, and was director of the Corfu psychiatric hospital from 1884 to 1887; Michael Katsaras (1860–1939), assistant director from 1887 to 1894 and then consulting physician, who had studied in Athens and Paris and later became the first professor of neurology and psychiatry in Greece; and Michael Yanniris (1865–1956), assistant director from 1894 to 1903 and then director until 1942, who had studied in Athens and France.[28] The medical director oversaw the medical organisation and functions of the institution, while the board of directors, made up not only of doctors but

[25] Michalis Yanniris, "Η επιστημονική δραστηριότητα του Δρομοκαΐτειου στα πρώτα πενήντα χρόνια της λειτουργίας του" ["The scientific activity of the Dromokaiteio in the first fifty years of its operation"], in Tsikis (ed.), *Dromokaitio Hospital, 1887–1987 + 15*, pp. 95–103.

[26] Konstantinos Pappis, "Η διοίκηση του Δρομοκαΐτειου" ["The administration of the Dromokaitio"], in Tsikis (ed.), *Dromokaitio Hospital, 1887–1987 + 15*, pp. 223–42; Karamanolakis, "The emergence of institutions", p. 81.

[27] *Λογοδοσία του Διοικητικού Συμβουλίου του εν Δαφνίω Φρενοκομείου Δρομοκαϊτείου, 1898: Πεπραγμένα [Report of the Board of Directors of the Dromokaitio Lunatic Asylum in Daphni, 1898: Activities]*, Athens: s.n., 1899, p. 10.

[28] Yanniris, "The scientific activity of the Dromokaitio"; Konstantinos Pappis, "Η διοικητική διεύθυνση του Δρομοκαΐτειου" ["The administrative directorate of the Dromokaitio"], in Tsikis (ed.), *Dromokaitio Hospital, 1887–1987 + 15*, pp. 243–50; Karamanolakis, "The emergence of institutions", p. 25.

also of businessmen and lawyers, had managerial and financial responsibilities, but also a say in who was admitted to the institution.[29] Finally, the Dromokaitio was subject to the control and supervision of the public authorities as provided for in law 742, and its director was appointed by the Minister of the Interior on the proposal of the board of directors and the opinion of the Medical Council.[30] There were therefore several different agents—public and private, medical and lay—at work at the Dromokaitio at the same time.

In 1904, another institution was added to the mental health constellation of Athens and Greece: the Eginitio Hospital. This was a university clinic founded on the initiative of Michael Katsaras, the first professor of psychiatry and neurology at the University of Athens (from 1893) and the first director of the Eginitio until 1931, when he became professor emeritus.[31] As a member of the committee that drew up in 1899–1900 the plans and budget for the clinic that was to be established with the bequest of Dionysios Eginitis, Katsaras persuaded the rectorate to devote the new hospital to the clinical teaching of "nervous diseases".[32] His role was also instrumental in shaping the physiognomy of the Eginitio as a hospital for the short-term hospitalisation of acute cases rather than for the long-term

[29] "On the approval of the provisional Regulation of the established Lunatic Asylum of the Venerable Zorzis Dromokaitis"; Pappis, "The administration of the Dromokaitio"; Karamanolakis, "The emergence of institutions", p. 62.

[30] "On the approval of the provisional Regulation of the established Lunatic Asylum of the Venerable Zorzis Dromokaitis". As in many countries, the Medical Council in Greece was an advisory body on matters of public health and medicine. It was established by royal decree in 1834 and was under the Ministry of the Interior.

[31] Katsaras was removed from his post and replaced by N. Pezopoulos and N. Makkas during the academic years 1910–1911 and 1911–1912, respectively. Neurological Clinic, University of Athens, *Αιγινήτειον Νοσοκομείον 1905–2005 [Eginitio Hospital 1905–2005]*, Athens: s.n., 2005, p. 16.

[32] National University, *Τα κατά την Πρυτανείαν Σπυρ. Π. Λάμπρου Πρυτανεύσαντος κατά το Ακαδημαϊκόν Έτος 1904–1905 [The Activities of the rectorate of Spyr. P. Lambros Rector during the Academic Year 1904–1905]*, Athens: Τύποις Π.Δ. Σακελλαρίου [P.D. Sakellarios Publishing House], 1907, p. 13; *Τα κατά την Πρυτανείαν Γεωργίου Ν. Χατζιδάκι, Τακτικού Καθηγητού της Γλωσσολογίας, Πρυτανεύσαντος κατά το Ακαδημαϊκόν Έτος 1905–1906 [The Activities of the Rectorate of Yeorgios N. Hadjidakis, Professor of Linguistics, Rector during the Academic Year 1905–1906]*, Athens: Τύποις Π.Δ. Σακελλαρίου [P.D. Sakellarios Publishing House], 1907. p. 38; Neurological Clinic, University of Athens, *Eginitio Hospital*, p. 8.

care of chronic patients (see Chapter 12).[33] Katsaras had studied medicine at the University of Athens, trained in neurology and psychiatry in Paris, at the Salpêtrière and the Sainte Anne hospitals, respectively, in 1880–83, and then, as we have seen, worked at the Dromokaitio.[34]

In the 1910s, the Eginitio had 96 beds and 150 patients per year,[35] who, as those in the Dromokaitio, came from various parts of Greece (see Chapter 7). Its purpose was twofold: the treatment of patients "suffering from nervous diseases and other diseases directly related to nervous diseases" and the clinical teaching of medical students, which was based on pauper patients.[36] The emphasis on nervous diseases had several origins. On the one hand, it was related to the scientific interests of Katsaras, whose doctoral thesis was on multiple sclerosis, but also to his view that in the absence of other neurological clinics, many "neuropaths" would come to the Eginitio to be treated as private, fee-paying patients, thus guaranteeing the hospital's viability.[37] On the other hand, neurology, with its organic connotations, did not carry the stigma of psychiatry and was a better advertisement for the institution, promoting the image of the Eginitio as a hospital not unlike those for physical illnesses. Besides, the nervous diagnoses during this period included conditions with psychological causes and symptoms and were sometimes used as euphemisms for mental illness (see Chapters 1 and 8). In any case, mental patients, diagnosed with typical psychiatric disorders of the time, such as "melancholia" and "mania", were admitted to the Eginitio from the beginning and

[33] Ploumpidis, *History of Psychiatry*, p. 209.

[34] Neurological Clinic, University of Athens, *Eginitio Hospital*, p. 16. Katsaras' move from the Dromokaitio to the Eginitio, as well as Tsirigotis' move from the Corfu hospital, where he was director from 1876 to 1887, to the Dromokaitio, shows that the few neurologists-psychiatrists of the time transferred between the main psychiatric institutions of the country.

[35] Libert, "Les aliénés en Orient", p. 298.

[36] "Περί εγκρίσεως Οργανισμού του Αιγινήτειου Νοσοκομείου" ["On the approval of the Statute of the Eginitio Hospital"], *Εφημερίς της Κυβερνήσεως [Government Gazette]*, 169, 30/7/1904. An advertisement in the press stated that the hospital treated "sufferers from various nervous diseases and from all kinds of pathological conditions with nervous phenomena requiring specialised treatment, such as indigestion, rheumatism, diabetes, migraines and arthritis in general, etc.". Anon., "Αιγινήτειον νοσοκομείον" ["The Eginitio hospital"], *Σκριπ [Skrip]*, 15/10/1904, p. 3.

[37] Proceedings of the rectorate's meeting, 11/12/1899, in Neurological Clinic, University of Athens, *Eginitio Hospital*, pp. 12–13.

made up the majority of patients admitted in the years under study (see Chapter 8). Furthermore, at least from the academic year 1907–1908, the Eginitio was designated as a neurological and psychiatric clinic.[38]

The hospital was not subject to law 742 and was governed by university authorities.[39] Its director was the professor of "Neurology and Phrenic Diseases" at the Medical School of the University of Athens. He was also a member of the hospital's board of administration along with three other members: the rector and two professors from the medical school, who were appointed each year by the senate. The rector, who was only coincidentally a doctor, represented the Eginitio "before the Courts and public authorities and in all its civil transactions"[40] and was involved in the policy of admitting patients (see Chapter 12). Thus, in the administration of the Eginitio, as in the Dromokaitio, medical and non-medical authorities coexisted, but only within the university.

In summary, psychiatric legislation and institutional care in Greece were part of the broader psychiatric reform movement in the Western world, which was designed as humanitarian and scientific, but at the same time focused on dealing with "dangerous" lunatics, maintaining public order and controlling public expenditure on the care of the mentally ill. In Greece, the results of this reform were meagre until the early decades of the twentieth century. Although hospital provision increased, it remained inadequate. Institutions were scarce, with the main deficits in public mental hospitals. Against this background, private philanthropy attempted to supplement state action, as it had done in other areas of health and welfare since the second half of the nineteenth century. An organised system of mental health care was not achieved, but rather a few isolated institutions were founded, operating at a national level and treating private, public and pauper patients. Each institution had its own character, its own regime of governance, and its own relationship with the state and the psychiatric law. With regard to the two institutions studied here, the Dromokaitio, although private, operated under law 742 and was subject to local and state authorities, while the Eginitio, a university hospital, was not subject to this law and was controlled by university authorities, mainly, though not exclusively, medical.

38 Neurological Clinic, University of Athens, *Eginitio Hospital*, p. 13.

39 Makris, *The Insane Criminals in Greece and Elsewhere*, p. 26.

40 "On the approval of the Statute of the Eginitio Hospital".

The limited expansion of psychiatric services in early twentieth-century Greece was related to the social and economic conditions of the time and, above all, to the degree of urbanisation and industrialisation. These two processes, as noted in Chapter 1, were major driving forces behind the emergence and spread of psychiatric institutions, in the sense that, by changing the dynamics within families and communities, they limited their ability to provide care.[41] In Greece, as discussed in Chapter 2, public order and public health problems intensified with the growth of cities in the late nineteenth and early twentieth century, leading to greater concern and intervention by the upper social classes and the authorities, and leading to the first initiatives for the development of psychiatric institutions. However, the country remained largely rural and traditional, and the family continued to play a central role in supporting and caring for its members. In this context, the network of psychiatric institutions remained limited. Although, as will be shown in the following chapters, there was a social demand for institutional mental health care in the early twentieth century—a demand that the existing hospitals could not meet—the establishment of psychiatric services was not one of the political and social priorities of the time.[42] It was not until the 1920s and especially in the 1930s, with the increasing urban development and economic and social upheaval, that state intervention in health and welfare, including the organisation of public psychiatric hospitals, became more widespread.[43]

Until then, the Eginitio and the Dromokaitio were two of the basic components of mental health care in Greece. They were very different from each other, the former being a university hospital and the latter a private asylum that also had a public function. However, they had one important feature in common: the multiplicity of actors that shaped their

[41] Andrew Scull, *Social Order/Mental Disorder: Anglo-American Psychiatry in Historical Perspective*, London: Routledge, 1989, pp. 213–38.

[42] Indicatively, every year some applications for admission to the Dromokaitio were rejected, as there usually more applications than vacancies. *Λογοδοσία του Διοικητικού Συμβουλίου του εν Δαφνίω Θεραπευτηρίου Ζωρζή και Ταρσής Δρομοκαΐτου κατά το Έτος 1921 [Report of the Board of Directors of the Hospital of Zorzis and Tarsi Dromokaitis in Dafni during the Year 1921]*, Athens: s.n., 1922, p. 5.

[43] A major force behind the economic and social changes that were directly linked to the growing organisation of state welfare and health care was the vast number of refugees arriving to Greece in the early twentieth century and mainly after the Greco-Turkish War of 1919–1921. These developments, however, are beyond the chronological scope of this study.

operation. On the one hand, as has already been highlighted, medical and lay authorities (local, state, university or charitable) played a role in the administration, admission policy and supervision of the institutions. On the other hand, as will be shown in the following chapters, the operation of both the Eginitio and the Dromokaitio was also shaped by other actors, namely the patients and members of their private and public environment, who sought access to the institutions. A first insight into these actors is offered in the following chapter in the light of the economic aspects of hospitalisation.

CHAPTER 4

The Economic Aspects of Institutional Mental Health Care

As mentioned in the previous chapter, the patients of the Eginitio and the Dromokaitio were both private, whose hospital fees were covered by their own resources, and paupers, whose fees were covered by public and private agents. The financial aspects of hospitalisation—how much it costed and who covered this cost—had a significant impact on who could receive institutional care and shaped the image of each institution as either private, with more or less well-off patients, or as charitable or public, with economically disadvantaged patients.

In the Eginitio and the Dromokaitio, patients were divided into three classes reflecting differences "in residence and variety of food".[1] In the Eginitio, first-class patients were placed in single rooms, second-class patients in rooms with one other patient and third-class patients in rooms with several other patients.[2] In the Dromokaitio, according to a 1914

[1] "Περί εγκρίσεως του προσωρινού Κανονισμού του συσταθέντος Φρενοκομείου του αοιδίμου Ζωρζή Δρομοκαΐτου" ["On the approval of the provisional Regulation of the established Lunatic Asylum of the Venerable Zorzis Dromokaitis"], *Εφημερίς της Κυβερνήσεως [Government Gazette]*, 192, 16/7/1887, article 76.

[2] Anon., "Αιγινήτειον νοσοκομείον" ["The Eginitio hospital"], *Σκριπ [Skrip]*, 15/10/1904, p. 3. However, the hospital registers record first- and second-class patients staying in shared rooms and third-class patients in single rooms. This shows that accommodation was not only determined by the height of the fees, but also by the availability of rooms

D. Kritsotaki, *A Social History of Mental Health Care in Greece, c. 1900–1920*, Mental Health in Historical Perspective,
https://doi.org/10.1007/978-3-032-05764-8_4

brochure, first-class patients had a private apartment with three rooms and personal service, while the patients of the second class had a private room but took their meals in a communal dining room.[3] The emphasis on luxurious services and the omission of any reference to the conditions in the third class illustrates that the hospital's advertising was aimed at wealthy patients, who would bring in more funds for the institution. Indeed, hospital fees were a necessary source of income for both the Dromokaitio and the Eginitio.[4]

Regarding the height of hospital fees, in the Eginitio hospitalisation costed 360, 210 and 120 drachmas per month for the first, second and third classes, respectively. From 1917, due to the financial problems of the hospital, the fees began to increase and in 1920 they reached 750, 510 and 360 drachmas per month for each class.[5] In the Dromokaitio, in 1887 the hospital fees were set at 350, 200 and 60 drachmas per month depending on the class, but at the beginning of the twentieth century they increased, reaching 450, 300 and 270 drachmas per month in 1920. The following year, due to the overpricing of goods and increases in staff salaries, the fees for the first two classes rose again to 600 and 400 drachmas per month, respectively.[6] Although hospitalisation was less expensive in the Dromokaitio than in the Eginitio, the fees for both, even in the third class, were high in relation to the average wages of the time. For example, unskilled industrial workers earned between 90 and 150

and the condition of the patients, since "agitated patients" were sometimes placed in single rooms.

[3] In Nikolaos Tsikis (ed.), *Δρομοκαΐτειο Θεραπευτήριο, 1887–1987 + 15. Εκατό Δεκαπέντε Χρόνια Κοινωνικής Προσφοράς [Dromokaitio Sanatorium, 1887–1987 + 15. One Hundred Fifteen Years of Social Offer]*, Athens: s.n., 2001, pp. 57–59.

[4] Anon., "Αι πανεπιστημιακαί κλινικαί και οι φοιτηταί της Ιατρικής" ["University clinics and medical students"], *Σκριπ [Skrip]*, 5/12/1909, p. 2; *Report of the Board of Directors of the Hospital of Zorzis and Tarsi Dromokaitis in Dafni during the Year 1921*, Athens: s.n., 1922, p. 11.

[5] *Α΄ Μητρώον Εσωτερικών Ασθενών Αιγινήτειου Νοσοκομείου* [*First Register of Inpatients of the Eginitio Hospital]* and *Γ΄ Μητρώον Εσωτερικών Ασθενών Αιγινήτειου Νοσοκομείου* [*Third Register of Inpatients of the Eginitio Hospital*], archive of the Eginitio Hospital, Eg/1 and Eg/1.1, Athens (henceforth Eg/1 and Eg/1.1).

[6] "On the approval of the provisional Regulation of the established Lunatic Asylum of the Venerable Zorzis Dromokaitis"; *Report of the Board of Directors of the Hospital of Zorzis and Tarsi Dromokaitis in Dafni during the Year 1921]*, pp. 4, 6.

Table 4.1 Class of patients admitted to the Eginitio (according to the fees paid)

Class	*Patients*	
	Number	*%*
First	151	10,6
Second	450	31,5
Third	607	42,4
Reduced fees	26	1,8
Pauper	187	13,1
Unknown	9	0,6
Total	1430	100,0

Source: Sample from the Registers, Eginitio, 1905–1909 and 1916–1920 (see Table 1.1).

drachmas per month in the early 1910s.[7] Fee-paying hospitalisation thus presupposed that the patients or their environment had a certain financial comfort, while the two upper classes were clearly reserved for patients of a higher economic status.

These patients appear to have been few. In the Eginitio, where the class of patients was recorded in the registers, it can be ascertained that most of the patients were treated in the third class, and only a minority in the first (Table 4.1). It is interesting to note here that psychiatric patients had a higher concentration in the first two classes than neurological patients. This implies that despite the emphasis placed on neurology by the hospital's medical authorities, explained in Chapter 3, the hospital was more financially reliant on psychiatric patients. The latter generated more income for the institution, not only because they were more frequently admitted to the higher classes, but also because they were more numerous and stayed for longer periods in the institution than neurological patients (see Chapters 8 and 12).

The inference that most of the patients were not well-off is confirmed by the problems of several to pay for their hospital fees. During the period under consideration, according to the patient registers, 24 patients were discharged from the Dromokaitio for late payment, while in the Eginitio there were 38 patients whose fees were overdue, 23 of whom appear to

[7] Christina Agriantoni, "Βιομηχανία" ["Industry"], in Christos Hadziiossif (ed.), *Ιστορία της Ελλάδας του 20ού Αιώνα [History of Greece in the 20th Century]*, v. A1, Athens: Βιβλιόραμα [Vivliorama], 1998, pp. 173–221.

have been discharged for this reason.[8] In addition, in the Eginitio 97 patients were transferred to a lower class due to late payment, usually one or two months after admission, which suggests a rapid reduction in the financial resources available to these patients or a change in the hospitalisation strategy: as long as it was hoped that the patients would improve quickly, the available financial resources were used generously; as the first months passed and it became clear that the patients needed longer hospitalisation, the resources available to them were cut back in order to last longer. In any case, long-term hospitalisation could not be financed by such temporary means as borrowing or pawning. The example of Aristotle S. is typical: he had saved the money for the first month and a half of his wife's treatment in the Eginitio in the third class, "by pawning what precious jewellery I had, since I can barely survive with my large family of small children on my meagre salary of one hundred and twenty-four drachmas". This "poor civil servant", as he described himself, appealed to the "charitable sentiments" of the university senate so that his wife could be treated free of charge.[9]

The relatives of several other patients of the Eginitio asked to be exempted from paying the hospital fees or to pay with a discount or credit, stressing that they were poor breadwinners, or that they faced financial difficulties because of the condition of the patients and the cost of hospitalisation.[10] The wife of a third-class patient, whose fees were paid for the most part by the municipality of Athens and for one-sixth by herself, asked to be exempted from her share of 20 drachmas on the grounds that she had no income, since her husband was the breadwinner and she had no other "family support".[11] Although these letters may be somewhat exaggerated or even inaccurate, they demonstrate that private psychiatric hospitalisation was financially demanding and disrupted the family's financial balance.

[8] As stated on the back of the receipts issued by the Eginitio, a delay of more than three days in the payment of hospital fees gave the institution the right to discharge the patient to his/her guardian. Such receipts can be found in the folder *Κλινικές-Νοσοκομεία [Clinics-Hospitals]*, 1912–1913, Archive of the Rectorate of the University of Athens, Historical Archive of the University of Athens.

[9] Letter dated 10/7/1910, *Clinics-Hospitals*, 1909–1910.

[10] Letter dated 5/9/1920, *Clinics-Hospitals*, 1919–1920.

[11] Letter dated 16/10/1915, *Clinics-Hospitals*, 1915–1916.

Patients who were not even marginally able to be treated at their own expense could apply for free treatment at the expense of the institution, a benefactor, a charity, or a public agent. Law 742 (article 35) stipulated that, while the cost of treatment for the mentally ill was normally borne by themselves or their families, in cases of "pauperism" it was to be borne by the patients' municipality or, if the municipality was unable to provide the necessary financial assistance, by the state. The local administration and the state were therefore obliged to finance the psychiatric hospitalisation of indigent patients by paying part or all of the hospital fees, usually in the third class. In addition, the Eginitio and the Dromokaitio covered the costs of a number of patients from the private patients' fees and from other sources specific to each institution: in the Eginitio from the fees of outpatients undergoing hydrotherapy and electrotherapy, and in the Dromokaitio from donations and bequests, such as the original bequest of Zorzis Dromokaitis, the donation for the maintenance of pauper patients from Chios made by the businessman and banker Andreas Syggros in 1889, and the 1899 Syggros bequest.[12] Philanthropy, then, was officially instituted at the Dromokaitio, but, as we shall see below, it was at work in both hospitals. The balance between hospitalisation at private expense and that at the expense of the institutions, philanthropy or the public sector can shed light on whether the institutions operated as private, charitable or public institutions and can thus contribute to our understanding of the organisation of institutional health care in Greece.

As shown in Table 4.1, the Eginitio had only a few "paupers". These were patients who were treated at the institution's expense and were selected mainly on the basis of clinical criteria (see Chapter 12). Besides, their number was limited by the predetermined ten beds that could be occupied by the "indigent".[13] A few patients who paid reduced third-class fees (60–100 drachmas per month), which were supplemented by the institution, were also recorded in the registers as "paupers". However, apart from the "paupers", there were patients who were hospitalised for a fee, but did not pay this out of their own resources. Although we do not know the exact extent of this type of hospitalisation, the documents in the

[12] Anon., "The Eginitio Hospital"; Vangelis Karamanolakis, "Η ανάδυση των ιδρυματικών θεσμών. Το Δρομοκαΐτειο φρενοκομείο (1880–1903)" ["The emergence of institutions. The Dromokaitio Lunatic Asylum (1880–1903)"], MA thesis, Athens, National and Kapodistrian University of Athens, 1997, p. 30.

[13] Letter dated 19/10/1915, *Clinics-Hospitals*, 1915–1916.

university and hospital archives record public (state and local) and private agents who paid the patients' fees. Some agents were regular: the Ministry of the Military, the Ministry of the Interior and the Ministry of Welfare, as well as organisations operating under the auspices of ministries, such as the Committee for the Relief of Refugees and the Patriotic League of Greek Women, which later became the Patriotic Foundation of Welfare. Other agents were not as stable: prefectures, municipalities and communities, the University of Athens, the Ionian Academy, embassies, banks and associations such as the Cretan Brotherhood in Athens. Overall, public agents played a greater role than private ones. For example, the Ministry of Welfare, according to its own records, paid for 131 patients from 1917, when it was founded, up to 1920;[14] the Ministry of the Military had special arrangements for the admission of soldiers and officers: from 1911 it had arranged for the admission of non-commissioned officers in payment of their salaries, namely 60–70 drachmas per month, and during the Balkan Wars (1912–1913) it had commandeered 20 second- and third-class beds for officers and soldiers, respectively.[15] Soldiers continued to be admitted to the Eginitio on special terms after the Balkan Wars, and during the Great War several soldiers and officers were hospitalised there.[16]

Public admissions were met with opposition from the medical authorities of the Eginitio, especially during periods of financial hardship for the institution, such as the 1910s, because public patients' fees were often lower than normal or were paid belatedly. In 1917, for example, the Ministry of the Military owned the Eginitio 27,364 drachmas.[17] It was in this context that the medical director reacted to public hospitalisation. Indicatively, in 1911, he wrote to the rector that the treatment of non-commissioned officers for 60–70 drachmas per month could not be continued indefinitely for financial reasons and lack of space.[18]

[14] Michael Ailianos, *Το Έργον της Ελληνικής Περιθάλψεως [The Work of Greek Welfare]*, Athens: s.n., 1921, pp. 409–13.

[15] Letter dated 26/5/1911, *Clinics-Hospitals*, 1910–1911; Letter dated 26/5/1913, *Clinics-Hospitals*, 1913–1914.

[16] Letter dated 18/4/1918, *Clinics-Hospitals*, 1917–1918.

[17] Letter dated 18/11/1917, *Clinics-Hospitals*, 1917–1918.

[18] Letter dated, 26/5/1911, *Clinics-Hospitals*, 1910–1911.

In the Dromokaitio, the registers did not record whether patients were treated at their own expense or at the expense of the institution or other agents. Some information is provided by the annual reports of the hospital, which document that at the end of the period studied private patients made up the smallest part of its population. The decline of the paying clientele of the Dromokaitio may be linked to the establishment of private clinics in Athens at the early twentieth century, which were favoured by the upper social classes.[19] In any case, in 1921 most patients were paupers, supported by the institution or the public (Table 4.2). The hospitalisation of indigent patients at the hospital's expense began in the year of its foundation and was promoted by the board of directors as a stable charitable function meant to compensate for the shortage of mental hospitals and to relieve the poor mentally ill, their families and society in general. However, the institution could not cope with the numerous requests for free treatment. For this reason, the board of directors appealed to the state and private sponsors for financial assistance.[20] At the same time, it selected who would receive free treatment, mainly on the basis of social criteria, such as the financial situation of the patients, and their place of origin, since there was a special donation for patients from Chios. Moreover, the endorsement of the application by an important person could facilitate its acceptance, and this made the procedure subject to non-meritocratic criteria.

Initially, the number of patients maintained by the Dromokaitio was small (only six in 1888), but by the beginning of the twentieth century it had increased, reaching 100 (27% of the total patients) in 1915.[21] In 1921, they were fewer, but now public patients had increased from 25% in 1915 to over 67%, becoming the majority of patients (Table 4.2). In the same year, the main sponsor of public patients was the Ministry of the Interior, namely the state authority responsible for the Dromokaitio. Other important sponsors were the Ministry of the Military, which in the late 1910s had commandeered 30 beds for "psychopathic soldiers"

[19] Dimitris Ploumpidis, *Ιστορία της Ψυχιατρικής στην Ελλάδα. Θεσμοί, Ιδρύματα και Κοινωνικό Πλαίσιο 1850- 1920 [History of Psychiatry in Greece: Institutions, Hospitals and Social Framework 1850–1920]*, Athens: Εξάντας [Exandas], 1995, pp. 193–94.

[20] Karamanolakis, "The emergence of institutions", p. 63.

[21] Ibid., p. 30 and Ploumpidis, *History of Psychiatry*, p. 203, table 47.

Table 4.2 Payment of fees in the Dromokaitio (1921)

Payment	*Patients*	
	Number	*%*
"At their own expense" (private patients)	79	21,2
"At the expense of the institution"	42	11,3
By the Ministry of the Interior	164	44,1
By the Ministry of the Military	46	12,4
By the Ministry of Navy	3	0,8
By the Ministry of Welfare	5	1,3
By the Patriotic Foundation of Welfare	2	0,5
By the municipality of Athens	16	4,3
By the municipality of Piraeus	15	4,0
(Total of public patients)	(271)	(67,5)
Total of patients treated in 1921	372	100,0

Source: *Λογοδοσία του Διοικητικού Συμβουλίου του εν Δαφνίω Θεραπευτηρίου Ζωρζή και Ταρσής Δρομοκαΐτου κατά το Έτος 1921 [Report of the Board of Directors of the Hospital of Zorzis and Tarsi Dromokaitis in Dafni during Year 1921]*, Athens: s.n., 1922, pp. 5–6.

and 10 for "psychopathic officers",[22] the municipalities of Athens and Piraeus, each with 20 beds,[23] and the Ministry of Welfare, which paid for 128 patients for a total of 3,349 days between 1918 and the first half of 1921.[24] As in the Eginitio, public sponsors were often unreliable in their payments. The municipality of Athens had been the institution's largest debtor since the end of the nineteenth century.[25] In fact, at the beginning of the twentieth century, the Dromokaitio tried—without success—to pause admitting patients of the municipality until the latter had settled its debts.[26]

In conclusion, psychiatric hospitalisation was funded by patients, their private environment, the public, as well as by the hospitals themselves and

[22] Michael Yanniris, [Letter to the Prefect], 1918, *Α΄ Βιβλίο εξ Επαγγέλματος Εισαγομένων Ψυχοπαθών [First Book of ex Officio Admitted Psychopaths]*, archive of the Dromokaitio Hospital, p. 89.

[23] Ploumpidis, *History of Psychiatry*, p. 190.

[24] Ailianos, *The Work of Greek Welfare*, pp. 409–13.

[25] Karamanolakis, "The emergence of institutions", p. 31.

[26] Anon., "Αι οφειλαί του Δήμου προς το Δρομοκαΐτειον" ["The debts of the Municipality to the Dromokaitio"], *Σκριπ [Skrip]*, 11/5/1904, p. 2.

charitable agents. In both the Eginitio and the Dromokaitio, all these sources of funding were present but in different proportions, reflecting the different character of the two institutions. The Eginitio operated primarily as a private and university institution, while the Dromokaitio operated as a private and philanthropic institution, which at the same time had a public profile, as it was subjected to law 742. At the end of the period under consideration, the Dromokaitio's public function had intensified: its public patients had increased, while its private patients and those who were treated at the hospital's expense had decreased. Its authorities, like those of the Eginitio, resented public hospitalisation and tried to control it, because it meant lower fees and debts, but the Dromokaitio, exactly because it operated under law 742, was less likely to limit the number of public patients than the Eginitio. Another difference between the two hospitals was that the Eginitio selected the patients it financed according to clinical criteria that served its scientific objectives (see Chapter 12), whereas the Dromokaitio's criteria were social, dictated by its charitable mission.

Yet, both institutions tried to attract private patients in order to increase their revenues. The extent of hospitalisation "at private expense" shows that there was a demand for paid hospitalisation in Greece at the time, which was accessible mainly to the middle and upper social strata, as the hospital fees were quite high. However, most private patients were treated in the lowest class, and some found it difficult to keep up with the costs. As a result, affluent patients, although they existed, were a small part of the population of the institutions. At the same time, there were demands for free and low-cost psychiatric care that were not being met. On the one hand, the financial problems of both hospitals determined a policy of restricting the admission of pauper patients. On the other hand, the Dromokaitio was always full, while the Eginitio was not obliged to accept public patients. Thus, in the early twentieth century institutional mental health care was hard to access for the disadvantaged social strata, as there were neither suitable and adequate public institutions nor stable and merit-based mechanisms for administering free hospitalisation. The Dromokaitio and the Eginitio were neither intended nor equipped to make up for these shortcomings.

PART II

The Patients of the Institutions

In both the Eginitio and the Dromokaitio, as in most psychiatric hospitals of their time around the world, a number of characteristics were recorded for each patient, including gender, age, marital status, occupation, place of residence, diagnosis and symptoms of illness. Furthermore, psychiatrists were interested in and explored the ways in which social characteristics were interrelated with mental illness and its symptoms. The following chapters analyse the patients' characteristics and the psychiatrists' discourses on them, not to trace the demography of mental illness but to discuss the social (Chapters 5–7) and medical (Chapter 8) physiognomy of the institutions and their populations, and to highlight the social connotations of medical discourses. This part therefore contributes both to the mapping of psychiatric theory and to the analysis of the social functions of psychiatry and institutional mental health care, a theme that will be further developed in the third and final part of the book.

CHAPTER 5

Men and Women

Psychiatrists in the nineteenth and early twentieth centuries commonly argued that the causes and forms of mental illness differed between men and women (see Chapter 8), and they also inquired whether the two sexes were equally affected by mental disorders. This chapter focuses on the latter point, exploring psychiatric discourses on the comparative psychiatric morbidity of men and women. It then analyses, through the examination of admissions to the Eginitio and the Dromokaitio, the ways in which gender intertwined with psychiatric hospitalisation and the use of institutions.

Gender and Mental Illness

When investigating who suffered more from mental illness, men or women, psychiatrists usually used statistics from mental hospitals, which were the main reservoir of data on mental illness in the nineteenth and early twentieth centuries. Since this data varied between institutions and countries, there were psychiatrists who claimed that men and women were equally affected by mental illness, and others that one sex was more

D. Kritsotaki, *A Social History of Mental Health Care in Greece, c. 1900–1920*, Mental Health in Historical Perspective,
https://doi.org/10.1007/978-3-032-05764-8_5

affected than the other.[1] Psychiatrists explained the differences in statistics and opinions by arguing that the ratio of male to female insane varied across countries, seasons, habits, and lifestyles. Indicatively, the relationship between insanity and gender was seen as varying between urban and rural populations, and between upper and lower social classes.[2] Some psychiatrists argued that the quantitative differences between female and male mentally ill were not substantial and that if there were more female insane, it was because women outnumbered men in the general population and male lunatics had a higher mortality rate than women as they were more likely to suffer from fatal conditions, such as general paralysis of the insane.[3]

The diverse views on which sex was more vulnerable to mental illness were largely expressed in biological terms.[4] Those who supported that women were more prone to mental disorders justified this on the basis of the smaller size of their skull and therefore of their brain,[5] and on the

[1] By the mid-nineteenth century, all three views had been proposed. See the analysis on sex in two seminal psychiatric books: Jean-Étienne Dominique Esquirol, *Des Maladies Mentales: Considérées sous les Rapports Médical, Hygiénique et Médico-Légal*, v. 1, Paris: J.-B. Baillière, 1858, pp. 34–38; John Charles Bucknill and Daniel Hack Tuke, *A Manual of Psychological Medicine: Containing the History, Nosology, Description, Statistics, Diagnosis, Pathology, and Treatment of Insanity: With an Appendix of Cases*, London: J. Churchill, 1858, pp. 244–47. For an outline of the relevant views up to the end of the nineteenth century, see Havelock Ellis, "Sex, influence of, in insanity", in Daniel Hack Tuke (ed.), *A Dictionary of Psychological Medicine: Giving the Definition, Etymology and Synonyms of the Terms used in Medical Psychology with the Symptoms, Treatment, and Pathology of Insanity and the Law of Lunacy in Great Britain and Ireland*, v. 2, London: J. & A. Churchill, 1892, pp. 1152–56.

[2] Esquirol, *Des Maladies Mentales*, pp. 35–38; Bucknill and Tuke, *A Manual of Psychological Medicine*, p. 246; Ellis, "Sex, influence of, in insanity".

[3] For example, Maurice Craig, *Psychological Medicine. A Manual of Mental Diseases for Practitioners and Students*, 3rd edition, London: J. Churchill, 1917, pp. 34–35. On the general paralysis of the insane, see Chapter 8.

[4] Nineteenth- and early twentieth-century medicine perceived the intellectual, mental and social differences between the sexes largely on the basis of biological data, for example about the digestive, musculoskeletal, reproductive and nervous systems. These explanations were adopted and promoted by other experts, such as pedagogues. Eleni Fournaraki, *Εκπαίδευση και Αγωγή των Κοριτσιών: Ελληνικοί Προβληματισμοί (1830–1910): Ένα Ανθολόγιο [Schooling and Education of Girls: Greek Reflections (1830–1910): An Anthology]*, Athens: Γενική Γραμματεία Νέας Γενιάς [General Secretariat for Youth], 1987, pp. 31–33.

[5] William Hammond, *A Treatise on Insanity in its Medical Relations*, New York, 1883, pp. 101–4.

grounds that the female nervous system was less resilient than the male one and that women's nervous and mental organisation was more delicate and complex, so that women were by nature unable to cope with the pressures of mental illness as successfully as men.[6] But the main explanation of women's alleged nervous and mental sensitivity was linked to their reproductive system. In both sexes, reproductive activity was considered as exhausting the energy of the organism and predisposing to mental illness, but in each sex the dangers were attributed to different causes: in men to excessive sexual intercourse and masturbation, which were framed not as biological functions but as abuses, while in women precisely to their specific biological functions: menstruation, pregnancy and childbirth (especially in women who had many and closely spaced deliveries).[7] Menstruation and pregnancy in particular were associated with a state of general weakness and imbalance, with various nervous symptoms, as well as with impulses, perversions and even criminal acts.[8]

Such biological theories reflected and disguised social assumptions within the context of the established gender roles and served to maintain women's proper place in the home (see Chapter 2). But psychiatrists also explicitly expressed social beliefs on the character, abilities and roles

[6] Thomas Clouston, *The Hygiene of Mind*, London: Methuen & Co, 1906, p. 208.

[7] Charles Arthur Mercier, *A Text-book of Insanity and Other Mental Diseases*, London: George Allen and Unwin, 1914, pp. 16–18.

[8] Alexander Morison, "Outlines of lectures on mental diseases (1824)", in Vieda Skultans, *Madness and Morals: Ideas on Insanity in the Nineteenth Century*, London: Routledge & Kegan Paul, 1975, pp. 32–33; Henry Maudsley, *The Physiology and Pathology of the Mind*, London: Routledge/ Thoemmes Press, 1993 (first edition in 1867), pp. 207–8; Mercier, *A Text-book of Insanity*, pp. 16–21; Michael Katsaras, *Παθολογία των Νεύρων και Ψυχιατρική. Νόσοι του Εγκεφάλου, Νόσοι της Παρεγκεφαλίδος, Φρενίτιδες Νόσοι [Pathology of the Nerves and Psychiatry. Diseases of the Brain, Diseases of the Cerebellum, Phrenic Diseases]*, v. 2, Athens: Εστία [Estia], 1898, p. 398. On the significance of menstruation in medical discourse see also George Man Burrows, "Commnetaries on insanity" (1828), in Skultans, *Madness and Morals*, pp. 223–27; Henry Maudsley, "Body and mind: An inquiry into their connection and mutual influence, especially in reference to mental disorders" (1873), in Skultans, *Madness and Morals*, pp. 230–35. Menstruation was generally defined by late nineteenth- and early twentieth-century medicine and psychiatry as a disability, a disease and a threat to mental stability, although in the early twentieth century a new generation of women doctors began to challenge these perceptions. Julie-Marie Strange, "Menstrual fictions: Languages of medicine and menstruation, c. 1850–1930", *Women's History Review*, 9 (2000), pp. 607–28; Julie-Marie Strange, "The assault on ignorance: Teaching menstrual etiquette in England, c. 1920s to 1960s", *Social History of Medicine*, 14 (2001), pp. 247–65.

of men and women.[9] For example, some argued that women were less vulnerable to insanity than men not only because of their nature, but also because the privacy of their lives and their limited public roles made them less exposed to the causes of insanity.[10] At the same time, with regard to upper-class women, mental illness was attributed to a lack of practical interests and an excess of religiosity,[11] to an absence of proper exercise and useful education[12] and to an "abnormal" lifestyle with activities that excessively and unnaturally stimulated their sensitive nervous systems, such as many and long social events.[13]

From another perspective, too much focus on marriage and the family was also seen as a source of mental vulnerability. The English psychiatrist Henry Maudsley (1835–1918) argued that women who were disappointed in their marriages or who failed to marry were often driven to mental illness because they were dependent on men and had only marriage as a respectable prospect; under the existing social system they were deprived of the outlets that men enjoyed in active employment and sexual activity out of wedlock.[14] Furthermore, according to the Scottish psychiatrist Thomas Clouston (1840–1915), civilised society posed a significant risk to women's mental health: it could lead to nervous exhaustion because it forced women to be constantly alert in order to suppress certain instincts, especially sexual ones.[15] Within this context, some psychiatrists viewed positively the easing of restrictions on women's

[9] For an early review of the crucial role played by the medical profession in defining social and sexual roles, see John Haller and Robin Haller, *The Physician and Sexuality in Victorian America*, Urbana: University of Illinois Press, 1974.

[10] Bucknill and Tuke, *A Manual of Psychological Medicine*, pp. 246–47; Esquirol, *Des Maladies Mentales*, p. 35.

[11] Clouston, *The Hygiene of Mind*, pp. 211–12.

[12] Esquirol, *Des Maladies Mentales*, p. 35; Maudsley, *The Physiology and Pathology of the Mind*, p. 209.

[13] Esquirol, *Des Maladies Mentales*, p. 35; Hammond, *A Treatise on Insanity*, p. 104; "Anon., "Η εκπαίδευσις των κορασίων και η κληρονομικότης (κατά τον Γκυγιώ)" ["Girls' education and heredity (according to Guyau)"] (1901), in Fournaraki, *Schooling and Education of Girls*, pp. 483–86.

[14] Maudsley, *The Physiology and Pathology of the Mind*, pp. 202–3.

[15] Clouston, *The Hygiene of Mind*, p. 208.

lives from the late nineteenth century, particularly the expansion of women's education and work opportunities.[16]

However, psychiatrists, like other medical professionals, were unable or unwilling to escape the assumptions about gender biology that underpinned the validity of their scientific theories and social beliefs, and which in turn supported the established gender inequality. They were therefore in no way in favour of the equality of men and women in education, employment and life in general.[17] In the psychiatric discourse of the second half of the nineteenth century, gender equality was commonly portrayed as both dangerous and impossible due to women's physical, mental and emotional traits, which were perceived as diametrically opposed to those of men.[18] Indicatively, Maudsley, who, as noted above, exposed the impasses into which the social system led women, believed that women could not participate in the male world of higher education, work and politics because their bodies and minds were more or less diseased and incapable of hard work for a quarter of every month of the best years of their lives.[19]

[16] Maudsley, *The Physiology and Pathology of the Mind*, p. 209.

[17] On the medical discourses on gender inequality grounded on biology, anatomy, menstruation, the characteristics of ova and sperm and the digestive system, see Strange, "Menstrual fictions", pp. 607–28; John Starrett Hughes, "The madness of separate spheres: Insanity and masculinity in Victorian Alabama", in Mark Carnes and Clyde Griffen (eds), *Meanings for Manhood: Constructions of Masculinity in Victorian America*, Chicago: University of Chicago Press, 1990, pp. 53-66; Eleni Varika, *Η Εξέγερση των Κυριών. Η Γένεση μιας Φεμινιστικής Συνείδησης στην Ελλάδα, 1933-1907 [The Rising of the Ladies. The Genesis of a Feminist Consciousness in Greece, 1933–1907]*, Athens: Κατάρτι [Katarti], 2004, pp. 86-87; Fournaraki, *Schooling and Education of Girls*, pp. 32–35.

[18] Women were perceived as the opposite or the reverse of men as early as the eighteenth century. Thomas Laqueur, *Making Sex: Body and Gender from the Greeks to Freud*, Cambridge, MA: Harvard University Press, 1992, pp. 5–6.; Lesley Hall, "Interpreting abnormal psychology in the late nineteenth century: William James's spiritual crisis", in Waltraud Ernst (ed.), *Histories of the Normal and the Abnormal: Social and Cultural Histories of Norms and Normativity*, London: Routledge, 2006, pp. 183–204; Ornella Moscucci, *The Science of Woman: Gynaecology and Gender in England, 1800–1929*, Cambridge: Cambridge University Press, 1990, pp. 7–41. The views prevailing in the second half of the nineteenth century on the intellectual and emotional differences between the sexes are summarised in Granville Stanley Hall, *Adolescence: Its Psychology and Its Relations to Physiology, Anthropology, Sociology, Sex, Crime, Religion, and Education*, v. 2, London: S. Appleton, 1904, pp. 561–647.

[19] In Strange, "The assault on ignorance", p. 250.

As the nineteenth century drew to a close, women in many countries were demanding and gaining employment and educational rights, and their entry into the public sphere was slowly becoming a reality. Nevertheless, it remained reprehensible for much of the male-dominated psychiatric profession, which still propagated the ideology of separated spheres and pathologised gender non-conformity.[20] According to the American neurologist William Alexander Hammond (1828–1900), the "fashion" of giving men and women the same education was absurd, because it overloaded the female mind with subjects it could not grasp, such as mathematics.[21] Another widespread view was that, as women became increasingly involved in the struggle for survival, they were being exposed to pathogenic influences that had previously only affected men,[22] and were being pressured in ways that they could not cope with.[23] Beyond increasing the likelihood of mental illness, the changes in women's lives and roles had another serious consequence, according to many medical authorities of the time: the inability to fulfil maternal duties, either because women who were educated and worked as men neglected their female responsibilities, or because the waste of girls' energy on education during the critical period of puberty exhausted them and endangered their biological maturation and fertility.[24] Such a development was regarded as dangerous for the family, the nation, the race—as eugenic ideas gained ground—but also for women themselves, because the lack of a family was seen as causing them mental strain.[25] The women's question was just one instance of psychiatrists offering their "expert" opinion. They did the same on various political events, social

[20] Jessie Hewitt, *Institutionalizing Gender. Madness, the Family, and Psychiatric Power in Nineteenth-Century France*, Ithaca: Cornell University Press, 2020.

[21] Hammond, *A Treatise on Insanity*, p. 104.

[22] Ellis, "Sex, influence of, in insanity".

[23] Clouston, *The Hygiene of Mind*, p. 211.

[24] Thomas Clouston, *Clinical Lectures on Mental Diseases*, 5th edition, London: J. & A. Churchill, 1898, p. 580; Hall, *Adolescence*, pp. 561–647. On similar medical opinions in Greece see Fournaraki, *Schooling and Education of Girls*, pp. 36–72; Varika, *The Rising of the Ladies*, pp. 134–36.

[25] Clouston, *The Hygiene of Mind*, p. 209.

phenomena and ideological issues, such as the anti-slavery movement in America and the anarchist movement in turn-of-the-century France.[26]

Greek psychiatrists of the late nineteenth and early twentieth centuries also perceived women's entry into the public sphere as incompatible with womanhood, harmful to women and society, and contrary to Greek customs and traditions.[27] However, they were reassured that most Greek women still led restrained lives, centred on the home and family, and not too involved in public life, thus keeping safe from pathogenic factors such as financial and professional problems, political concerns, rivalries, excessive and exhausting intellectual work, sexual and alcoholic abuse.[28] This, at least, was how Greek doctors attempted to explain the trend, which had also been observed by foreign psychiatrists, that women in Greece were less frequently admitted to psychiatric institutions than in other European countries.[29] Nevertheless, as the twentieth century progressed, the difference between Greece and European countries narrowed. By the early

[26] Lynn Gamwell and Nancy Tomes, *Madness in America: Cultural and Medical Perceptions of Mental Illness before 1914*, Binghamton: Cornell University Press, 1995, pp. 81–82; Edward Erickson, "Punishing the mad bomber: Questions of moral responsibility in the trials of French anarchist terrorists, 1886–1897", *French History*, 22 (2008), pp. 51–73.

[27] Katsaras, *Pathology of the Nerves and Psychiatry*, v. 2, p. 615; Michael Yanniris, "Περί των εξ οινοπνεύματος παθήσεων εν Ελλάδι" ["On diseases caused by alcohol in Greece"], *Ιατρική Πρόοδος [Medical Progress]*, 8 (1903), pp. 179–83; Simonidis Vlavianos, "Ο αλκοολισμός" ["Alcoholism"], *Ψυχιατρική και Νευρολογική Επιθεώρησις [Psychiatric and Neurological Review]*, 8 (1909–1910), pp. 3–192. Historians have pointed out that at the end of the nineteenth century, women's entry into the paid workforce was criticised as being linked to immorality and the negative influence that Europe was thought to have on Greek women. Efi Avdela, *Δημόσιοι Υπάλληλοι Γένους Θηλυκού. Καταμερισμός της Εργασίας κατά Φύλα στον Δημόσιο Τομέα, 1908–1955 [Female Public Servants: Labour Division by Gender in the Public Sector, 1908–1955]*, Athens: Ίδρυμα Έρευνας και Παιδείας της Εμπορικής Τράπεζας της Ελλάδος [Research and Education Foundation of the Commercial Bank of Greece], 1990, p. 26.

[28] Michael Yanniris, *Αι Φρενοπάθειαι εν Ελλάδι: Μελέτη επί Στατιστικών και Κλινικών Παρατηρήσεων [The Phrenopathies in Greece: A Study on Statistical and Clinical Observations]*, Syros: s.n., 1898, pp. 5–6; Vangelis Karamanolakis, "Η ανάδυση των ιδρυματικών θεσμών. Το Δρομοκαΐτειο φρενοκομείο (1880–1903)" ["The emergence of institutions. The Dromokaitio Lunatic Asylum (1880–1903)"], MA thesis, Athens, National and Kapodistrian University of Athens, 1997, p. 41.

[29] Esquirol, *Des Maladies Mentales*, p. 34; Dromokaitio Hospital, *Απολογισμός και Στατιστική του Δρομοκαϊτείου Φρενοκομείου Έτους 1889 [Report and Statistic of the Dromokaitio Lunatic Asylum for the year 1889]*, Athens: s.n., 1890, p. 35.

1930s, it was claimed that Greek men and women were equally affected by mentally illness.[30]

Gender and Psychiatric Hospitalisation

Although asylum statistics were used by psychiatrists to draw conclusions about mental illness, they speak mostly about the patients admitted to mental hospitals.[31] Thus, the ratio of men to women in institutions reveals less about the relative psychiatric morbidity of the sexes than about the gendered use of psychiatric institutions. Both the Eginitio and the Dromokaitio admitted more men than women—a trend consistent with quantitative analyses of asylum admissions in the nineteenth and early twentieth centuries in other countries—but the sex difference was more pronounced in the Dromokaitio than in the Eginitio (Diagram 5.1).[32] The ratio of male to female admissions is 2.56 in the former and 1.58 in the latter, and it is not reversed even if the army and navy patients admitted during the First World War and the Greco-Turkish War are excluded. Nor can the sex ratio be explained by the general demographic trends of the time, as men and women were roughly equal in the general population.[33] Other factors, therefore, determined the ratio of male to female patients in the Eginitio and the Dromokaitio.

[30] Konstantinos Mitaftsis, *Ψυχιατρική μετά 54 Προσωπογραφιών [Psychiatry with 54 Portraits]*, Athens: s.n., 1933, p. 20.

[31] In other words, it is a mistake to confuse hospital admission with morbidity. As social science research has shown, the use of mental health services does not depend solely on pathology, but on a number of factors, such as previous experience of services and the existence of alternative care options and support networks. Anne Rogers and David Pilgrim, *A Sociology of Mental Health and Illness*, 6th edition, London: Open University Press, 2021, pp. 74–75.

[32] Historical research on the proportion of men and women patients in mental hospitals in different countries has differentiated between patients admitted (more men) and patients in the hospital population (more women). See Joan Busfield, "The female malady? Men, women and madness in nineteenth century Britain", *Sociology*, 28 (1), 1994, pp. 259–77; Solange Lapeyrière, "Celles qui sortent et celles qui restent. 'Carrières asilaires' des femmes internées dans les asiles en France au XIX siècle", *Clio. Femmes, Genre, Histoire*, 51, 2020, pp. 283–308; Anatole Le Bras, "Troubles dans la masculinité", *Histoire, Médecine et Santé*, 23 (2023), pp. 55–69.

[33] *Στατιστικά Αποτελέσματα της Γενικής Απογραφής του Πληθυσμού κατά την 27 Οκτωβρίου 1907 [Statistical Results of the General Census of the Population on 27 October 1907]*, v. 1, Athens: Εθνικό Τυπογραφείο [National Press], 1909, p. νγ; *Απογραφή του Πληθυσμού*

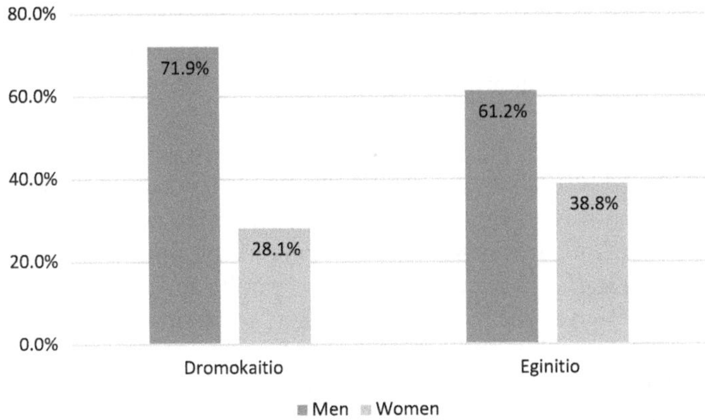

Diagram 5.1 Men and women admitted to the Dromokaitio and the Eginitio. Source: Samples from the registers, Dromokaitio and Eginitio (1905–1909 and 1916–1920). Total number of patients sampled 1430 and 961 respectively (see Table 1.1).

A first interpretation of the marked difference between male and female admissions is provided by the observation of Christodoulos Tsirigotis, director of the Dromokaitio from 1887 to 1903, that families were reluctant to admit their female members because psychiatric hospitalisation was believed damaging to marriage prospects.[34] In other words, admissions reflected and were affected by the willingness to use institutions, which in turn was related to socially accepted gender roles. The stigma of insanity and asylum confinement weighed differently—and more heavily—on women due to their destination to marry and have a family.

But the impact of gender roles on asylum admissions was broader. The lower number of women than men in the Eginitio and the Dromokaitio should be seen as a result and an indication of the reluctance and resistance to psychiatric hospitalisation of women than of men, largely due to

της Ελλάδος κατά την 19 Δεκεμβρίου 1920. Γενικά Στατιστικά Αποτελέσματα [Census of the Population of Greece on 19 December 1920. General Statistical Results], Athens: Εθνικό Τυπογραφείο [National Press], 1928, table 2, p. 11.

[34] Cited in Karamanolakis, "The emergence of institutions", p. 41.

the ideology of separate spheres and domesticity. This, while unresponsive to the complex realities of private and public life, exerted a powerful influence on gender identities, expectations and realities (see Chapter 2), including psychiatric hospitalisation. To understand how these ideals influenced hospital use, admission to psychiatric institutions can be seen as a move away from the private sphere: compared to the home, the psychiatric institution was a public space. Within its walls, patients were under the gaze of unfamiliar individuals: members of the medical, nursing, and administrative staff, other patients, and the public and medical authorities who were involved in the processes of patient admission and discharge and who were responsible for inspecting psychiatric institutions (see Chapters 3 and 11). The medical and non-medical control of admissions and hospitalisation was less intense in the case of private patients and especially in the Eginitio, where neither medical certification of patients nor disclosure of their data to the public authorities was required, as the hospital was not subject to law 742. On the contrary, all admissions to and discharges from the Dromokaitio had to be reported to the public authorities, and patients had to be examined and certified before admission by two city doctors, or by any two doctors sworn in before the Magistrates' Court.[35] In both hospitals, however, hospitalisation exposed patients and their relatives to the scrutiny of strangers—albeit to varying degrees—and meant that patients were literally and metaphorically removed from the private sphere of the home and the family.

Within the ideology of separate spheres, this departure was perceived as more difficult and damaging for women than for men, who were already seen as exposed in the public sphere. It is telling that other, more private loci of psychiatric treatment, such as small private clinics, attracted more female patients than psychiatric institutions. For example, the private clinic of the neurologist-psychiatrist Simonidis Vlavianos had more women than men in the 1910s, leading him to argue that there were more mentally ill women in Greece than was thought.[36]

[35] Informative leaflet of Dromokaitio, 1914, in Tsikis (ed.), *Dromokaitio Hospital, 1887–1987 + 15*, pp. 57–61; *Λογοδοσία του Διοικητικού Συμβουλίου του εν Δαφνίω Θεραπευτηρίου Ζωρζή και Ταρσής Δρομοκαΐτου κατά το Έτος 1921 [Report of the Board of Directors of the Hospital of Zorzis and Tarsi Dromokaitis in Dafni during the Year 1921]*, Athens: s.n., 1922, p. 4.

[36] At least this is what Vlavianos reportedly told the French doctor Lucien Libert during the latter's visit to Athens. Lucien Libert, "Les aliénés en Orient (Grèce, Empire Ottoman,

Yet, other spaces—the street, the city square and the neighbourhood—were more overtly public than the mental hospital. Psychiatric hospitalisation would not be regarded as so harmful for patients deemed strange, uncontrollable and dangerous in such spaces, causing embarrassment and upset to their private and public environment. Public conduct that was perceived as threatening, immoral or shameful was an integral part of psychiatric descriptions of mental illness (see Chapter 8) and was commonly identified as one of the reasons that prompted the patients' private and public environments to turn to psychiatric hospitalisation (see Chapters 9 and 10). More generally, unpredictable, violent and dangerous behaviour was rooted in official and popular conceptions of mental illness: psychiatric legislation in Greece, as in many countries, paid particular attention to "dangerous" lunatics,[37] while, as the administrators of the Dromokaitio in the late nineteenth and early twentieth centuries noted, "the people" saw lunatics mainly as violent and noisy[38] and psychiatric hospitals as "prisons".[39]

These connotations of mental illness and psychiatric institutions discouraged the psychiatric hospitalisation of women, especially those from the middle and upper classes. At the same time, the characteristics that made incarceration seem inevitable, notably uncontrolled and violent behaviour and the public manifestation of mental disorders, were more closely associated with men, who were generally perceived as more "difficult" patients than women. In contrast, women were thought to have quieter symptoms, to be easier to control and to keep out of the public

Égypte)", *L'Informateur des Aliénistes et des Neurologistes*, Supplement de l'*Encéphale*, 7 (1912), pp. 283–302.

37 See Chapter 3 and Mark Finnane, "From dangerous lunatic to human rights? The law and mental illness in Australian history", in Catherine Coleborne and Dolly MacKinnon (eds), *Madness in Australia: Histories, Heritage and the Asylum*, St. Lucia: University of Queensland Press, 2003, pp. 23–36.

38 Michael Yanniris, *Περί της Νοσηλείας και Περιθάλψεως των Φρενοπαθών εν τοις Φρενιατρείοις ή κατ' Οίκον: Προς Μόρφωσιν Νοσοκόμων [On the Care and Treatment of Insane Persons in Lunatic Asylums or at Home: For the Education of Nurses]*, Athens: s.n., 1903, pp. 90–91.

39 *Λογοδοσία του Διοικητικού Συμβουλίου του εν Δαφνίω Φρενοκομείου Δρομοκαϊτείου, 1898: Πεπραγμένα [Report of the Board of Directors of the Dromokaitio Lunatic Asylum in Daphni, 1898: Activities]*, Athens: s.n., 1899, p. 10.

space.[40] As a result, admission to psychiatric institutions to avoid exposure to more intensively public spaces was less likely to be pursued in the case of women than men.

Therefore, the ratio of men to women in the institutions can be largely attributed to the options and decisions of the patients' environment, which was less willing to admit women to mental hospitals and keener and more able to keep women at home. In addition, the higher proportion of women admitted to the Eginitio than to the Dromokaitio leads to the hypothesis that the type of institution also influenced who was admitted to it, as has been observed in other national contexts.[41] As noted above, public supervision was less pronounced in the Eginitio than in the Dromokaiteio, and the Eginitio had more private than public patients (see Chapter 4). These two features made the Eginitio look more like a private clinic than an asylum and can explain the higher proportion of women there than in the Dromokaiteio. Other characteristics of the

[40] For instance, according to Esquirol, men were "more maniacal" than women (Esquirol, *Des Maladies Mentales*, p. 38). On this gendered differentiation of mental symptoms, see Stephen Garton, "Asylum histories: Reconsidering Australia's lunatic past", in Coleborne and MacKinnon (eds), *Madness in Australia*, pp. 11–22; Le Bras, "Troubles dans la masculinité"; Akihito Suzuki, "Were asylums men's places?: Male excess in the asylum population in Japan in the early twentieth century", in Marijke Gijswijt-Hofstra, Harry Oosterhuis, Joost Vijselaar and Hugh Freeman (eds), *Psychiatric Cultures Compared: Psychiatry and Mental Health Care in the Twentieth Century. Comparisons and Approaches*, Amsterdam: Amsterdam University Press, 2006, pp. 295–311 (302–304); Michael MacDonald, *Mystical Bedlam: Madness, Anxiety, and Healing in Seventeenth-Century England*, Cambridge: Cambridge University Press, 1981, Chapter 3. The association of men's conduct with public manifestations of violent and aggressive behaviour is a trend also highlighted by social science research (Rogers and Pilgrim, *A Sociology of Mental Health and Illness*, pp. 70, 81, 85), while the rarer designation of women as dangerous and violent than men is not historically related solely to insanity: by the end of the eighteenth century the majority of recorded offences were committed by men, and female criminal behaviour took less violent forms than that of men. These gendered aspects of criminality have been explained in a similar way as those of mental illness: it has been argued that the sharpening of gender differences and the identification of women's ideal position in the private sphere meant that women had fewer opportunities to engage in criminal behaviour and that, when they did, they were more subject to informal, private systems of control rather than the legal system. Malcolm Feeley and Deborah Little, "The vanishing female: The decline of women in the criminal process, 1687–1912", in Robert Weiss (ed.), *The Social History of Crime, Policing and Punishment*, Aldershot: Ashgate, 1999, pp. 267–305.

[41] Joan Busfield, *Men, Women, and Madness: Understanding Gender and Mental Disorder*, Basingstoke: Macmillan, 1996, pp. 123–25.

Eginitio, related to its distinctive profile as a university clinic, encouraged the admission of women: shorter lengths of stay (see Chapter 11) and its location near the centre of Athens made admission to the Eginitio seem like a temporary separation from the family rather than a long-term separation with an unlikely return. Finally, as a university hospital and one orientated to neurology, the Eginitio enjoyed high scientific prestige and approximated hospitals for physical illnesses.[42] These features, combined with the fact that the Eginitio was a new institution at the beginning of the twentieth century, lightened the stigma of hospitalisation and clearly distinguished it from the Dromokaitio, an older and more typical mental hospital, away from the city centre, with an established, largely negative image.

Conclusion

Gender analysis yields important conclusions about both medical and lay perceptions of mental illness and the attitudes towards psychiatric institutions. Psychiatric discourse distinguished mentally ill men from women, identified differences between them in terms of biological (or biologically sounding) and social factors, and perceived women in comparison and contrast to men. It is also clear that psychiatric theories, even biological ones, had social origins and implications, as they presupposed, adopted and supported established gender roles of their time. Of course, scientific discourses constitute always "cultural narratives in disguise", since even data belonging to the field of physiology and pathology are necessarily perceived through cultural perceptions.[43] More than that, though, psychiatrists outright defended or condemned various practices concerning the lifestyle, education and work of the sexes and explicitly rejected female emancipation. In other words, the social roles of the doctor, already evident in the late eighteenth century, when the asylum became "an instrument of moral uniformity",[44] tended to extend outside

[42] It was no coincidence that more than 40% of the women admitted to the Eginitio during the period studied were given neurological and nervous diagnoses. On the distinction between these two diagnostic types, and the gendered dimension of diagnoses, see Chapter 8.

[43] Laqueur, *Making Sex*, pp. 11, 236.

[44] Michel Foucault, *Madness and Civilization. A History of Insanity in the Age of Reason,* transl. Richard Howard, New York: Vintage, 1973, p. 259.

the mental hospital and beyond the treatment of "madness" towards a wider remit of defining the "normal" and controlling the "abnormal".[45]

Moving from mental illness to psychiatric hospitalisation and starting with the representation of women in psychiatric admissions, I argued that the extent and ways of using the institutions depended on the social perceptions and practices of the time, but also on the physiognomy of each institution. Conventional gender traits, determined by the ideology of separate spheres and domesticity, influenced the likelihood of hospitalisation. The association of mentally ill women with "quiet" manifestations of disorder that were (more) easily contained in the private sphere and of mentally ill men with uncontrollable and publicly displayed symptoms—associations that were common in medical and popular descriptions of mental illness in the nineteenth and early twentieth centuries—made hospitalisation seem less imperative and less appropriate for women. Although the family was not always the sphere of protection described by the domestic ideal, often involving violence and social control functions, the removal of women from the home was perceived—at least ideologically—as harmful. Admission to psychiatric hospitals was therefore avoided more for women than for men, especially as these hospitals were generally viewed with reluctance or suspicion by the public. In this context, the ratio of men to women in psychiatric admissions was also linked to the characteristics of the different institutions and their degree of social approval. There were fewer female admissions to the Dromokaitio, which was more like a typical lunatic asylum, than to the Eginitio, which, as a university clinic, had a more positive image and subjected its patients to less public exposure and shorter hospital stays.

[45] Michel Foucault, *Le Pouvoir Psychiatrique: Cours au Collège de France (1973–1974)*, ed. by J. Lagrange, Paris: Seuil/Gallimard, 2003, pp. 219–20; Michel Foucault, *Les anormaux: Cours au Collège de France Année 1974–1975*, Paris: Seuil/Gallimard, 1999, pp. 150–51.

CHAPTER 6

Young and Old, Married and Not

The examination of the sex balance of admissions to institutions constructed men and women as homogeneous groups. Sex, however, intersects with the other social characteristics of the patients: age and marital and social status. By analysing the medical discourse and the records of the Eginitio and the Dromokaitio, this chapter continues the endeavour to trace the medical and social perceptions of mental illness, the physiognomy of the patients of the institutions and, by extension, the social function and position of mental hospitals.

Age

Throughout the nineteenth and early twentieth centuries, age was an inherent category of psychiatric thinking, as it was considered that mental illness did not have the same characteristics, causes and incidence at all

D. Kritsotaki, *A Social History of Mental Health Care in Greece, c. 1900–1920*, Mental Health in Historical Perspective,
https://doi.org/10.1007/978-3-032-05764-8_6

ages.[1] The division of men's and women's lives into childhood, adolescence, adulthood, maturity, climacteric and old age played an important role in psychiatric theory, although not all doctors assigned these periods to exactly the same years. Childhood was usually placed up to the age of 13 and adolescence up to the age of 25.[2] The end of maturity and the beginning of the next age, which included the climacteric and old age, were even more complex issues, being defined in relation to sex: maturity was usually considered to end at 45,[3] 50[4] or 55 for men,[5] and at 40 for women,[6] and old age to begin, again depending on sex, between 50[7] and 65.[8]

The different psychiatric views of the beginning and end of each life stage, and more generally the varying perceptions of the stages of life in different societies and times, highlight that these distinctions are not biologically given but socially and culturally constructed.[9] For this reason,

[1] For example, Jean-Étienne Dominique Esquirol, *Des Maladies Mentales: Considérées sous les Rapports Médical, Hygiénique et Médico-Légal*, v. 1, Paris: J.-B. Baillière, 1858, p. 32; Henry Maudsley, *The Physiology and Pathology of the Mind*, London: Routledge/ Thoemmes Press, 1993 (first edition in 1867), p. 212; Emmanuel Régis, *Précis de Psychiatrie*, 5th edition, Paris: Octave Doin et fils, 1914, p. 32; Michael Katsaras, *Παθολογία των Νεύρων και Ψυχιατρική. Νόσοι του Εγκεφάλου, Νόσοι της Παρεγκεφαλίδος, Φρενίτιδες Νόσοι [Pathology of the Nerves and Psychiatry. Diseases of the Brain, Diseases of the Cerebellum, Phrenic Diseases]*, v. 2, Athens: Εστία [Estia], 1898.

[2] Thomas Clouston, *Clinical Lectures on Mental Diseases*, 5th edition, London: J. & A. Churchill, 1898, p. 585.

[3] Ministry of the Interior, *Στατιστική του εν Κερκύρα Φρενοκομείου του Έτους 1877 [Statistic of the Lunatic Asylum in Corfu in the Year 1877]*, Athens: Εθνικό Τυπογραφείο [National Press], 1878, p. 7.

[4] Maudsley, *The Physiology and Pathology of the Mind*, p. 212.

[5] Clouston, *Clinical Lectures*, pp. 617, 639.

[6] Maudsley, *The Physiology and Pathology of the Mind*, p. 212.

[7] Ibid., p. 212.

[8] Clouston, *Clinical Lectures*, pp. 617, 639.

[9] The concepts of childhood and adolescence have been central to the historical and sociological theorisation of life stages. A turning point in historiography was the work of Philippe Ariès, who in 1960 argued that in the Middle Ages and until the seventeenth century there was no clear boundary between childhood and adulthood, and that, consequently, there was no recognition of the characteristics that would define the age category of the "child" and distinguish it from that of the adult. Ariès interpreted adolescence in a similar way, claiming that it was a concept that took a long time to form. Philippe Ariès, *Centuries of Childhood*, New York: Vintage Books, 1960. Ariès' arguments have

the following analysis adopts a grouping of ages based on the psychiatric, social and legal standards of the time with childhood extending up to 13 years, adolescence up to 20 years and youth up to 25 years. Youth is used to describe the transitional period between adolescence and maturity, because the psychiatric categorisation of adolescence up to 25 years seems incompatible with both the social and the legal realities of the early twentieth century, when working and married life often began before 25 and adulthood was defined by law at 21.[10] Finally, maturity is defined as covering the years from 26 to 39, middle age from 40 to 59 and old age from 60.

While the general population of Greece at that time was dominated by the younger age groups, with children accounting for 34% and adolescents for 11% of the population,[11] most patients admitted to the institutions were of mature and middle age (Diagram 6.1). The average age of admission was 32.9 years in the Eginitio and 33.6 years in the Dromokaitio. In the latter patients were concentrated in the 26–39 age group, while in the Eginitio they were spread over a wider age range: compared to the Dromokaitio more children, adolescents and middle-aged and elderly patients were admitted. However, children, the most

been widely criticised; see, for example, Julia Grant, "Parent-child relations in Western Europe and North America, 1500-present", in Paula S. Fass (ed.), *The Routledge History of Childhood in the Western World*, Routledge, London, 2013, pp. 103-24. Before Ariès, the sociologist Norbert Elias associated the differentiation of children's behaviour and make-up from that of adults with the course of civilisation. Norbert Elias, *The Civilizing Process: The History of Manners*, New York: Urizen Books, 1978 (first edition in 1939). Sociology, and especially since the 1980s the "new sociology of childhood", has produced remarkable research and insights into the meanings of childhood. For a review see, Alisson James, Chris Jenks and Alisson Prout, *Theorizing Childhood*, London: Polity, 2010. Anthropology has also clearly demonstrated that childhood is not understood in a single, universal way across the world and across civilisations. Heather Montgomery, *An Introduction to Childhood: Anthropological Perspectives on Children's Lives*, Chichester, West Sussex, U.K.: Wiley-Blackwell, 2009.

10 Kostas Sofianos, *Το Νομικό Καθεστώς της Παιδικής Ηλικίας και της Νεότητας 1833–1900: Συναγωγή Νόμων, Διαταγμάτων, Εγκυκλίων, Καταστατικών [The Legal Status of Childhood and Youth 1833–1900: A Compilation of Laws, Decrees, Circulars, Statutes]*, Athens: Γενική Γραμματεία Νέας Γενιάς [General Secretariat for Youth], 1988, p. 27. Sofianos used similar age groups: childhood up to 12 years, adolescence from 13 to 19 years and youth from 19 to 25 years.

11 *Απογραφή του Πληθυσμού της Ελλάδος κατά την 19 Δεκεμβρίου 1920. Γενικά Στατιστικά Αποτελέσματα [Census of the Population of Greece on 19 December 1920. General Statistical Results]*, Athens: Εθνικό Τυπογραφείο [National Press], 1928, table 2, p. 10.

numerous age group in the country's population, were almost absent from both hospitals.

These trends were in line with the psychiatric theory of the time, which, after all, as we saw in the previous chapter, was largely based on the observation of patients in institutions. The director of the Corfu psychiatric hospital stressed that the admissions to the institution in 1877 confirmed the "axiom" accepted by all asylum doctors, namely that "the two poles of life are less subject to insanity than adolescence and manhood", adding that "insanity develops rapidly from the age of 20–45".[12] This phenomenon was explained by the prevailing psychiatric theory that mental illness was more frequent and more intense when the organism, especially the brain and nervous system, was in full

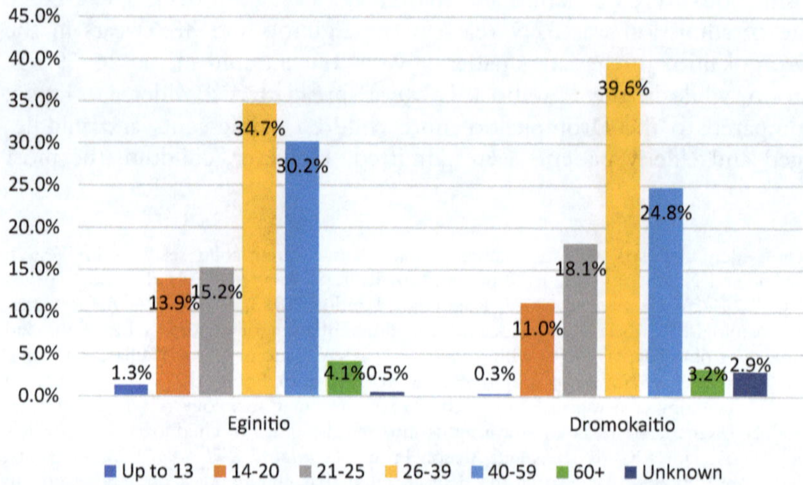

Diagram 6.1 Ages of the patients admitted to the Eginitio and the Dromokaitio. Source: Samples from the registers, Eginitio and Dromokaitio, 1905–1909 and 1916–1920 (see Table 1.1). The youngest child admitted was 8 years old in the Eginitio and 5 years old in the Dromokaitio

[12] Ministry of the Interior, *Statistic of the Lunatic Asylum in Corfu in the Year 1877*, p. 7.

development, which was reached at the age of maturity.[13] It was also suggested that mature people were more exposed to the causes of insanity, both physical (alcohol, syphilis, overwork) and moral, i.e. psychological (responsibilities, worries, emotional and mental stress).[14]

However, neither middle age nor old age was considered free of mental illness:[15] organic brain diseases could cause mental symptoms; significant biological changes were sources of mental vulnerability; and the moral consequences of withdrawal from active life and loss of interests contributed to increased incidence of mental breakdown.[16] Children, on the contrary, were deemed as more protected from mental illness. It was claimed that they were not threatened by the multitude of physical and moral factors that endangered older people, they rested and slept well, and their brains had significant nourishing capacities, so that any problems could be corrected by the organism without the children and those around them noticing.[17] At the same time, the child's brain and nervous system were understood as lacking the complexity that made mental illness probable and dangerous, as functions that were not developed were considered unlikely to be disrupted.[18]

If childhood was protected from mental illness, adolescence was considered to be the first essentially dangerous age for the onset of mental disorders.[19] As doctors observed a sharp increase in the number of admissions to mental hospitals after childhood, which was also the case in the two institutions in this study, they concluded that the incidence of mental illness increased in adolescence and sought to explain the relationship between adolescence and insanity in the biology and lifestyles

[13] Thomas Clifford Allbutt, "Children, insanity in", in Tuke (ed.), *A Dictionary of Psychological Medicine*, v. 1, pp. 202-5; Clouston, *Clinical Lectures*, p. 577; Maurice Craig, *Psychological Medicine. A Manual of Mental Diseases for Practitioners and Students*, 3rd edition, London: J. Churchill, 1917, p. 35.

[14] Bucknill and Tuke, *A Manual of Psychological Medicine*, p. 247; Charles Arthur Mercier, *A Text-book of Insanity and Other Mental Diseases*, London: George Allen and Unwin, 1914, p. 16.

[15] Maudsley, *The Physiology and Pathology of the Mind*, p. 212.

[16] Mercier, *A Text-book of Insanity*, pp. 20–21.

[17] Clouston, *Clinical Lectures*, p. 584.

[18] Allbutt, "Children, insanity in".

[19] Maudsley, *The Physiology and Pathology of the Mind*, p. 212; Clouston, *Clinical Lectures*, p. 583.

of adolescents. Biological changes during adolescence, especially those affecting the reproductive system, were thought to play a crucial role if there was a marked heredity.[20] This was referred to as "self-poisoning of the pubertal process".[21] Another aggravating factor in adolescence was considered the tendency to indulge in "abuses" that weakened the body.[22] These abuses usually implied sexually reprehensible behaviour, and doctors strongly urged children and adolescents to avoid masturbation and sexual activity at least until the end of adolescence.[23] In addition, adolescence was seen as an age of intense mental activity, another noted risk factor for disorder.[24]

The above theories highlight the basic traits that psychiatric discourse attributed to each age: childhood was understood as the age of simplicity and innocence, in need of protection; adolescence was associated with the onset of temptation and abuse, difficulty in self-control and mental strain; maturity was identified with both energy and fatigue, worries and immorality; finally, middle and old age were described as ages of withdrawal from action, loss of interests and decline. These psychiatric representations of age, like those of gender, were expressed in biological and social terms and presupposed and reinforced the social constructs of the different age stages. In fact, the fundamental divergence that the psychiatric discourse established between children and adults, that children belonged in the quiet and safe (private) sphere and adults in the dynamic but perilous (public) sphere, stemmed from the same ideology that regulated gender roles and relations: the ideology of separate spheres and domesticity.

Age roles and assumptions undoubtedly determined perceptions and practices of psychiatric hospitalisation, about which, as noted above, the data from institutions have more to say than about the prevalence of mental illness. Because maturity was more closely associated with the public sphere and an active lifestyle than the other ages, adult problems were more readily recognised and hospitalisation was more urgently

[20] Katsaras, *Pathology of the Nerves and Psychiatry*, v. 2, p. 578, which stated, for example, that in girls the disorders often started with the onset of menstruation.

[21] Régis, *Précis de Psychiatrie*, p. 376.

[22] Katsaras, *Pathology of the Nerves and Psychiatry*, v. 2, p. 578.

[23] Clouston, *Clinical Lectures*, pp. 135, 216.

[24] Ibid., p. 577.

sought, so that the patients could be cured or brought under control and returned to their duties.[25]

The opposite could be said for children, who were less readily hospitalised,[26] although the psychiatric expertise on childhood that developed from the second half of the nineteenth century argued that children had a higher percentage of mental illness than was generally thought.[27] These experts claimed that childhood mental health problems were kept hidden from medical observation, because children were rarely perceived as dangerous, and were generally considered easy to care for and control at home.[28] In addition, some parents did not want to acknowledge or did not understand the true nature of the problems and were reluctant to submit children to medical examination and treatment, relying instead on nurses and teachers even for children with severe mental disorders. The Greek neurologist-psychiatrist Simonidis Vlavianos noted that the "poor parents" of even epileptic children, were "flattered" to believe that they would do well in mainstream schools, when in fact they needed special education or hospital care.[29] The pedagogue Aristotle Kourtidis observed that "abnormal" children tended to be seen as simply ill-nurtured and badly behaved, because the family did not want to admit that they were mentally ill, but also because it was ignorant about mental illness. Furthermore, Kourtidis underlined that the already difficult distinction between mental health and illness was even more difficult in children, since "many

[25] Ellen Dwyer, *Homes for the Mad: Life Inside Two Nineteenth-Century Asylums*, New Brunswick, N.J.: Rutgers University Press, 1987, p. 92.

[26] For an intriguing account including an autobiographical narrative of a child (although of an older age) in an asylum, see Anatole Le Bras, *Un Enfant à l'Asile—Vie de Paul Taesch (1874–1914)*, Paris: CNRS Éditions, 2018.

[27] For example, Paul Moreau (de Tours), *La Folie chez les Enfants*, Paris: J.-B. Baillière, 1888; John Langdon Down, *On Some of the Mental Affections of Childhood and Youth*, London: J. & A. Churchill, 1887; M. B. Ray, "Case of acute mania occurring in a boy", *Journal of Mental Science*, 44 (1898), pp. 320–21; William Ireland, *The Mental Affections of Children: Idiocy, Imbecility and Insanity*, 2nd edition, London: J. & A. Churchill, 1900 (first edition in 1898).

[28] Allbutt, "Children, insanity in".

[29] Simonidis Vlavianos, "Τα μη κανονικά παιδία" ["The non-normal children"], *Ψυχιατρική και Νευρολογική Επιθεώρησις [Psychiatric and Neurological Review]*, 6 (1907–1908), pp. 11–13.

of the degenerate children keep their soul shut [...] they play like ordinary children, talk or act like other children".[30] After all, according to the psychiatrists who specialised in children, the very nature of childhood was characterised by such features as a strong imagination, an unclear boundary between the real and the unreal, bad temper and a tendency to disobedience,[31] which made it difficult to determine when children crossed the limits of what was tolerable for their age.[32]

The reluctance of the children's environment to see them as mentally ill and the tendency to keep them at home were consistent with the ideology of domesticity and the dominant ideals of childhood: children were at the centre of the family's unity and happiness and because of their presumed innocence were perceived as needing protection and control through their containment in the private sphere.[33] Against this background, the stigma of mental illness and the fear of asylums—especially as these were mostly for adult patients and associated with the "dangerously insane"—made psychiatric hospitalisation of children seem inappropriate or even cruel. It would have been easier to commit children to mental hospitals specialising in minors, but such hospitals did not exist in Greece at the time. The Eginitio, where there were slightly more children than in the Dromokaitio, may have seemed less inappropriate and threatening not only to women (see Chapter 5) but also to children because of its special physiognomy and more positive image as a newly established university hospital, not too far from the city centre, and with short stays. Furthermore, the fact that most of the children in the Eginitio were

[30] Aristotle Kourtidis, "Ψυχοπάθειαι εν τω σχολείω. Νύξεις προς τους διδασκάλους και τους γονείς" ["Psychopathies in the school. Suggestions to teachers and parents"], *Ψυχιατρική και Νευρολογική Επιθεώρησις [Psychiatric and Neurological Review]*, 2 (1903–1904), pp. 81–89. On degeneration, see Chapter 8.

[31] Allbutt, "Children, insanity in"; Fletcher Beach, "Insanity in children", *Journal of Mental Science*, 44 (1898), pp. 459–73; Thomas Clouston, *The Hygiene of Mind*, London: Methuen & Co, 1906, p. 121. Other scientific discourses also distinguished children's mental traits from those of adults. For example, Émile Durkheim noted that, in contrast to adults, the two main characteristics of the child were the irregularity (instability) of activities and mental manifestations and the lack of discipline and self-control. Émile Durkheim, *L'Education Morale*, Paris: Presses Universitaires de France, 1974 (first edition in 1925), pp. 110–13.

[32] Ireland, *The Mental Affections of Children*, p. 280.

[33] Chris Jenks, *Childhood*, London: Routledge, 2005, p. 86; Dimitra Makrinioti, *Η Παιδική Ηλικία στα Αναγνωστικά Βιβλία, 1834–1919 [Childhood in Literacy Books, 1834–1919]*, Athens/Ioannina: Δωδώνη [Dodoni], p. 38.

diagnosed with neurological disorders, while almost all the children in the Dromokaitio were diagnosed with mental retardation, suggests that neurological specialisation played a key role in the admission of children to the Eginitio. Neurological conditions were considered by doctors to be more common in children than mental illnesses and were easier to recognise, to accept and to lead to hospitalisation because they had physical symptoms and were perceived as organic diseases.[34] However, even in the Eginitio there were few children, which confirms that the psychiatric institutions studied were considered to be institutions for adults and were rarely used for children.

The number of elderly patients admitted to the Eginitio and the Dromokaitio was higher than that of children, but still relatively low. The lack of mental health care services in Greece may have discouraged institutions and the public sector from hospitalising older people, who were perceived as less likely to recover and thus "give back" to the family and society.[35] However, the low rate of admission of older people can be interpreted not only in terms of the shortcomings of the Greek psychiatric system, but also in terms of the care practices for the elderly. In general, during the nineteenth century and most of the twentieth century, care for the elderly in Greece was primarily a family matter and responsibility, and institutional alternatives such as workhouses and nursing homes were fewer than in other countries, such as Britain.[36] Like children, although probably to a lesser extent, the elderly were seen as

[34] William Hammond, *A Treatise on Insanity in its Medical Relations*, New York: D. Appleton and Co, 1883, p. 90.

[35] The reluctance to work with older people in mental health has been noted both now and in the past. See Claire Hilton, "Age inclusive services or separate old age and working age services? A historical analysis from the formative years of old age psychiatry c.1940–1989", *Psychiatric Bulletin*, 39: 2 (2015), pp. 90–95. The lukewarm response to mental health in old age extends to sociology, which has not focused on the issue until recently. See Paul Higgs and Chris Gilleard, "Ageing, dementia and the social mind: Past, present and future perspectives", *Sociology of Health & Illness*, 39: 2 (2017), pp. 175–81.

[36] Ioannis Mastroyannis, *Ιστορία της Κοινωνικής Προνοίας της Νεωτέρας Ελλάδος, 1821–1960 [History of Social Welfare in Modern Greece, 1821–1960]*, Athens: s.n., 1960, pp. 249, 287; Charitable Society of Athens, *Γηροκομείο 1864–1964 [Old People's Home 1864–1964]*, s.l.: s.n., s.d. On Britian, see Pat Thane, *Old Age in English History: Past Experiences, Present Issues*, Oxford: Oxford University Press, 2000, pp. 167–68, 288–98; Margaret Anne Crowther, "Poverty, health and welfare", in William Hamish Fraser and Robert John Morris (eds), *People and Society in Scotland, 1830–1914*, v. 2, Edinburgh: J. Donald, 1990, pp. 265–89.

vulnerable and attached to the home. The low numbers of admissions of older people to both institutions may therefore support the hypothesis that families were disinclined to use psychiatric institutions for members who were attributed an "organic" place in the home.

Marital Status

The marital status of patients was another piece of information recorded in the institutional archive and taken into account by psychiatrists when analysing the social components of mental illness, although it did not provoke the same degree of debate and disagreement as gender and age. In the nineteenth and early twentieth centuries, psychiatrists tended to agree that mental illness was more common among the unmarried,[37] which they explained in two ways. On the one hand, they argued that the mentally ill were less likely to marry[38] because of their "mental peculiarities",[39] "peculiar temperament" and "monotonous" and "selfish" lives.[40] On the other hand, the reverse was claimed, namely that marriage protected mental health because the freedom enjoyed by the unmarried deprived them of "moral principles" and encouraged "abnormal life".[41] In the early twentieth century, such formulations referred less to vague moral causes of mental illness and more to venereal diseases, especially syphilis, which was recognised as causing mental illness. Although this second way of linking celibacy with mental illness through sexual promiscuity mainly concerned men, women were also thought to be at particular risk of mental illness if they remained unmarried. Psychiatrists perceived

[37] Bucknill and Tuke, *A Manual of Psychological Medicine*, pp. 255-58. Michael Katsaras (*Pathology of the Nerves and Psychiatry*, v. 2, p. 398), in his analysis of melancholy, argued that suicide was more common among the unmarried than among the married and rare among those with children.

[38] Edward Renvoize and Allan Beveridge, "Mental illness and the late Victorians: A study of patients admitted to three asylums in York, 1880–1884", *Psychological Medicine*, 19 (1989), pp. 19–28.

[39] Bucknill and Tuke, *A Manual of Psychological Medicine*, p. 258.

[40] Konstantinos Mitaftsis, *Ψυχιατρική μετά 54 Προσωπογραφιών [Psychiatry with 54 Portraits]*, Athens: s.n., 1933, p. 21.

[41] Ibid. Although this source dates from the 1930s, similar views existed in the nineteenth century. See, for example, Bucknill and Tuke, *A Manual of Psychological Medicine*, p. 258.

marriage and family as a refuge, a life destination and a means of maintaining mental balance for women, and associated celibacy with a greater likelihood of employment, which, according to psychiatric theory at the time, exposed women to pathogenic pressures (see Chapter 5).

The data from the Eginitio and the Dromokaitio are consistent with the psychiatric views, as most patients admitted were unmarried. The married followed, while widowed and divorced patients were far fewer (Diagram 6.2). The latter two categories were very limited in the Greek population—there were few divorcees, and widowers were also much fewer than the married and unmarried.[42] With regard to widows, it should also be considered that the patients who were more likely to be widowed, i.e. the elderly, were fewer than the mature and middle-aged patients (Diagram 6.1). On the contrary, the ratio of unmarried to married patients in the institutions did not follow the general population trend of the time, when there were more married than unmarried people.[43]

However, as has been pointed out repeatedly, the characteristics of the inmates of the institutions say more about psychiatric hospitalisation and the ways in which the institutions were used than about mental morbidity. Thus, the higher proportion of unmarried patients in both hospitals indicates that the unmarried were hospitalised more often than the married, not that they were more vulnerable to mental illness. It seems, therefore, that having a spouse meant having a stronger family support network and was a factor that made it more likely to remain at home or at least outside psychiatric hospitals. This kind of association between marital status and hospitalisation is confirmed by the shorter length of stay of married than unmarried patients in both the Eginitio and the Dromokaitio. Not only were married patients admitted less frequently than unmarried patients, but they also stayed in the institutions for shorter periods of time: on average, married patients spent 147 days in the Eginitio and unmarried

[42] *Απογραφή του Πληθυσμού της Ελλάδος κατά την 19 Δεκεμβρίου 1920. Γενικά Στατιστικά Αποτελέσματα [Census of the Population of Greece on 19 December 1920. General Statistical Results]*, Athens: Εθνικό Τυπογραφείο [National Press], 1928, table 8, p. LXXXII; Evgenia Bournova, *Οι Κάτοικοι των Αθηνών (1900-1960) – Δημογραφία [The Residents of Athens (1900–1960)—Demography]*, Athens: Εθνικό και Καποδιστριακό Πανεπιστήμιο Αθηνών-Τμήμα Οικονομικών Επιστημών [National and Kapodistrian University of Athens-Department of Economics], 2016, pp. 136–41, 174–78.

[43] In 1920, 54% of the population aged 15 years and over were married, and 36% were unmarried. *Census of the Population of Greece on 19 December 1920*, table 8, p. LXXXII.

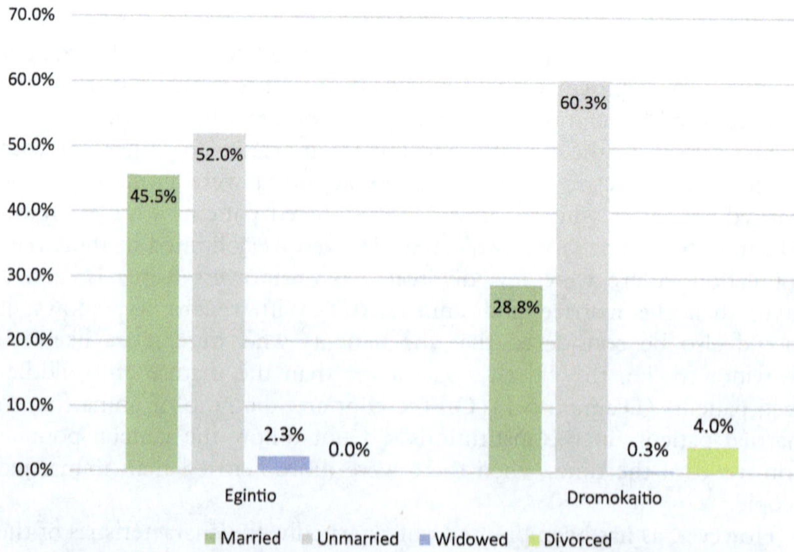

Diagram 6.2 Marital status of patients (aged 15 and over) admitted to the Eginitio and the Dromokaitio. Source: Samples from the registers, Eginitio and Dromokaitio, 1905-1909 and 1916-1920 (see Table 1.1). Patients under 15 years of age, who are expected to be unmarried, have been excluded. No marital status was recorded for 0.2% and 6.6% of patients admitted to the Eginitio and the Dromokaitio respectively

patients 209 days, while in the Dromokaitio the corresponding figures were 908 days and 1461 days. The higher proportion of married patients in the Eginitio than in the Dromokaitio is probably related to the generally shorter stays in the former: shorter-length hospitalisation was better suited to patients who could be cared for outside the institution, which points more to the married than to the unmarried.

Conclusion

This chapter has highlighted the ways in which psychiatric discourse incorporated and maintained socially accepted beliefs and practices about marriage, family life and age (childhood, adolescence, adulthood, middle age, and old age). At the same time, it has shown how marital status and age, with all their social and cultural connotations, influenced institutionalisation, in conjunction with the capacity and availability of mental health services.

This is particularly evident in the case of age. Mature patients, who were perceived to be more socially useful, more vulnerable to mental illness and more difficult to control, were given priority for hospitalisation, especially as the limited number of psychiatric beds imposed a selection of who was admitted. In contrast, the perception of all other age groups as economically inactive and of children in particular as easier to manage and less susceptible to mental illness made their admission to psychiatric institutions seem less necessary. At the same time, it was thought more appropriate for children and the elderly to remain at home, since removal from it was contrary to family ideals and practices. Especially for children, admission to an institution seemed threatening, bringing them into a place of adults, and, moreover, insane and potentially "dangerous" adults. In this context, even the neurological specialisation and the positive image of the Eginitio as a modern university clinic could not significantly increase the number of children hospitalised.

The analysis of admissions by age allows another set of conclusions to be drawn about the use of psychiatric institutions. Although the motives for admitting patients were not the same according to age, the examples of children and adults show that psychiatric hospitalisation could have many reasons and purposes, including the control, treatment and social reintegration of the patients. At any event, the fact that children, adolescents and the elderly, as was the case with women, were admitted in small numbers can be taken as an indication that mental hospitals were perceived as harsh institutions, in which groups considered vulnerable and in need of family protection, would not be able to cope. It therefore seems that in early twentieth-century Greece psychiatric institutions were viewed with distrust. The rural character of the country, the important place of the family in Greek society, and the recent appearance and limited expansion of psychiatric institutions did not favour their acceptance and unconditional use.

CHAPTER 7

Where Did Patients Live and Work?

Like gender, age and marital status, occupation and place of residence were key components in how mental illness was understood by nineteenth- and early twentieth-century psychiatrists. As social and cultural factors that shaped the conditions and meanings of patients' lives, they played a role in medical and lay perceptions of mental illness and in decisions about hospitalisation. In this chapter, I briefly examine the significance that work, employment and place of residence held in the psychiatric discourses before focusing on occupation and place of residence as indicators of both the social profile of patients and the reception of psychiatric institutions in early twentieth century Greece.

Occupation

Psychiatric theory of the nineteenth and early twentieth centuries frequently referred to work and occupation. Experts on mental illness stressed that excessive, ceaseless and sedentary work could cause mental breakdown, especially in people predisposed to mental illness.[1] Even

[1] Maurice Craig, *Psychological Medicine. A Manual of Mental Diseases for Practitioners and Students*, 3rd edition, London: J. Churchill, 1917, p. 31.

D. Kritsotaki, *A Social History of Mental Health Care in Greece, c. 1900–1920*, Mental Health in Historical Perspective,
https://doi.org/10.1007/978-3-032-05764-8_7

psychiatrists who attached less importance to the moral, i.e. psychological, causes of mental illness did not fail to mention work-related concerns as the most frequent moral cause of mental illness, especially in men.[2] Even in the case of women, psychiatric discourse linked mental disorder both to the lack of suitable employment and to the expansion of women's work from the second half of the nineteenth century and especially at the turn of the century (see Chapter 5). Moreover, certain occupations were considered more dangerous: those that were monotonous but also too risky and speculative, thus generating intense stress; those that involved excessive mental effort, but also exposure to the sun and harmful substances such as carbon[3]; finally, the occupations of wine merchant and tavern keeper, who tended to consume the spirits they sold, and of innkeeper, merchant and soldier, who usually led unstable lives and were therefore at risk of contracting syphilis.[4]

The idea that occupation had an effect on mental health extended beyond medical discourses. In the case histories from the Eginitio and the Dromokaitio, the patients and their environment commonly referred to work and specific occupations when attempting to explain mental illness. Physical exhaustion from work was mentioned in the cases of housewives,[5] labourers,[6] and farmers.[7] In other instances, moral strain from work was noted as a causal factor: a teacher was reported to have become irritable since she thought she had been dismissed,[8] and a civil servant was said to have "developed symptoms of nervousness" "as a consequence of

[2] Thomas Clouston, *Clinical Lectures on Mental Diseases*, 5th edition, London: J. & A. Churchill, 1898, p. 124.

[3] Alexander Morison, "Outlines of lectures on mental diseases (1824)", in Vieda Skultans, *Madness and Morals: Ideas on Insanity in the Nineteenth Century*, London: Routledge & Kegan Paul, 1975, pp. 32–33; Jean-Étienne Dominique Esquirol, *Des Maladies Mentales: Considérées sous les Rapports Médical, Hygiénique et Médico-Légal*, v. 1, Paris: J.-B. Baillière, 1858, pp. 40–46.

[4] Konstantinos Mitaftsis, *Ψυχιατρική μετά 54 Προσωπογραφιών* [*Psychiatry with 54 Portraits*], Athens: s.n., 1933, pp. 20–21.

[5] *Πρόχειραι Ιστορίαι Νόσου Εισαγομένων Ασθενών Β΄* [*Draft Illness Histories of Patients Admitted B*], archive of the Eginitio Hospital, 9.2, Athens, p. 353 (henceforth Eg/9.2).

[6] Eg/9.2, p. 265; *Βιβλίον Ιστορικών* [*Case Histories Book*], archive of the Dromokaitio Hospital, Athens (henceforth Dr/Case histories), 1918–1919, p. 1085.

[7] Eg/9.2, pp. 337, 423.

[8] Eg/9.2, p. 375.

not being promoted".[9] There were also references to the habit of grocers, waiters, brewers and farmers to drink alcohol at work.[10]

It is certainly possible that some of these details emerged in the lay narratives in response to specific questions from doctors, or that they were associated with the mental health problems by the doctor rather than by the informants or patients. However, it is not unlikely that the informants and patients themselves attributed the disorders to disappointments and fatigue from work, or to unemployment and subsequent poverty. Such problems were common lay explanations of mental illness, easy to identify and narrate, and corresponding to important aspects of the identity and life of men and women.[11] Although from the second half of the nineteenth century doctors, especially those who adhered to the biological model of mental illness, were sceptical of conventional psychological lay explanations of mental illness,[12] the patient histories of the two institutions show that there was no complete break between lay and medical perceptions at the beginning of the twentieth century: work-related factors, such as physical and moral strain, continued to be accepted by experts and non-experts alike. Doctors continued to rely heavily on evidence provided by patients and their families, and institutions remained a site of encounter and exchange between medical and lay perspectives. The persistence of experts and non-experts in linking mental disorders to work and occupation confirms the conclusion of the previous two chapters, namely that there was continuity and interaction between professional and lay conceptions of mental illness.

While occupation and work were categories through which doctors and lay people understood and explained mental illness, the occupations recorded in the patient registers, as well as the details of patients'

[9] Dr/Case Histories, 1919–1920, pp. 1213–14.

[10] Eg/9.2, pp. 341, 445.

[11] Marjorie Levine-Clark, "'Embarrassed circumstances': Gender, poverty, and insanity in the West Riding of England in the early Victorian years", in Jonathan Andrews and Anne Digby (eds), *Sex and Seclusion, Class and Custody: Perspectives on Gender and Class in the History of British and Irish Psychiatry*, Amsterdam: Rodopi, 2004, pp. 123–48.

[12] The "biological psychiatry" of the second half of the nineteenth century, represented mainly by Wilhelm Griesinger, Theodor Meynert and Carl Wernicke in Germany, Charles Lasègue and Joseph Babinski in France, and Henry Maudsley in Britain, emphasised the biological aspects of mental illness, i.e. the anatomy and chemistry of the brain and nervous system. Edward Shorter, *A History of Psychiatry: From the Era of the Asylum to the Age of Prozac*, New York: John Wiley, 1997, ch. 3.

work described in the case histories, can be used as indicators of the social physiognomy of patients and, ultimately, of the social reception of institutions, as has been done in the previous chapters for gender, age and marital status. To this end, the occupations recorded in the registers were grouped together due to their variety—indicatively, 286 different occupations were recorded for the 1430 patients in the Eginitio sample—by economic sector and, at a second level, by economic and social status (Table 7.1).[13] This grouping has several limitations, mainly because social classifications should be based on a combination of occupation with other economic, social and cultural indicators, such as additional economic activities and income, place of residence, standard of living, religion, education and ideological horizons,[14] information which is missing from the registers. Nevertheless, the internal differentiation of economic sectors is necessary to give some idea of the economic and social status of patients, as there are strong differentiations within each economic sector. For example, industry/manufacturing includes everyone from factory owners to workers.

The occupations in the first subgroup of each economic sector are associated with higher income and social prestige and, with all the reservations about the correspondence between occupation and class, they refer to the upper social strata and primarily to the upper and middle bourgeoisie.[15] The designation "landowner", without definitely corresponding to large

[13] For similar groupings of occupations in economic sectors, see Christina Agriantoni, *Οι Απαρχές της Εκβιομηχάνισης στην Ελλάδα τον 19ο Αιώνα* [*The Beginnings of Industrialisation in Greece in the 19th Century*], Athens: Εμπορική Τράπεζα της Ελλάδος [Commercial Bank of Greece], 1986, pp. 357–59; Yannis Yannitsiotis, *Η Κοινωνική Ιστορία του Πειραιά, 1860–1910. Η Συγκρότηση της Αστικής Τάξης* [*The Social History of Piraeus, 1860–1910. The Formation of the Urban Class*], Athens: Νεφέλη [Nefeli], 2006, p. 60; Walter Alan Armstrong, "The use of information about occupation", in Edward Anthony Wrigley (ed.), *Nineteenth-Century Society: Essays in the Use of Quantitative Methods for the Study of Social Data*, London: Cambridge University Press, 1972, pp. 191–310.

[14] Joanna Bourke, *Working-Class Cultures in Britain 1890–1960: Gender, Class and Ethnicity*, London: Routledge, 1994, pp. 2–3.

[15] For a definition of the middle classes and the bourgeoisie, which were broad categories, encompassing a variety of socio-economic levels, from the financial and intellectual elites to the numerous small artisans, traders and clerks, see Kostas Raptis, "Αστικές τάξεις και αστικότητα στην Ευρώπη, 1789–1914: Προσανατολισμοί της σύγχρονης ιστοριογραφίας" ["Middle classes and middle-class culture in Europe, 1789–1914: Approaches in modern historiography"], *Μνήμων* [*Mnimon*], 20 (1998), pp. 211–43; Stana Nenadic, "The Victorian middle classes", in William Hamish Fraser and Irene Maver (eds), *Glasgow, 1830–1912*, v. 2, Manchester: Manchester University Press, 1996, pp. 265–99.

Table 7.1 Grouping of patients' occupations

Economic sector	*Occupations*
Industry/ manufacturing	Industrialists, factory owners[16]
	Engineers
	Engineering assistants, electricians, ironmongers, mechanical engineers, skilled or semi-skilled artisans/producers (tailors, bakers, etc.)
	Workers
Building	Civil engineers
	Contractors
	Craftsmen (builders, carpenters, painters, plumbers, etc.)
	Workers
Mining	Engineers
	Supervisors and managers of mines
	–
	Workers
Agriculture, livestock, fisheries	Landowners[17]
	Agriculturalists
	Farmers, shepherds, fishermen, divers, gardeners
	Agricultural servants and farm workers
Public and private sector and armed forces	High-ranking officials of the armed forces and civil servants (e.g. majors, military doctors, members of parliament, judges)
	Senior employees of the private sector (e.g. directors and managers of banks and companies)
	Low-ranking officers of the armed forces and civil servants (e.g. sergeants, telegraph and postal employees, railway employees, policemen, clerks)
	Lower employees of the private sector (e.g. accountants and bank clerks)
	Clerks, porters, doormen, cleaners
	Sailors and soldiers[18]

(continued)

[16] Although in the nineteenth century the terms factory owner and industrialist could also mean artisan, craftsman and worker, by the end of the century they tended to take on their current meaning, namely that of a proprietor of a usually large industrial unit. Yannitsiotis, *The Social History of Piraeus, 1860–1910*, pp. 92–95.

[17] Those described as "landowners" in nineteenth-century Greece did not necessarily live in the countryside or work the land but could have other economic activities. Christos Loukos, *Πεθαίνοντας στη Σύρο τον 19ο αιώνα. Οι Μαρτυρίες των Διαθηκών* [*Dying in Syros in the nineteenth century. The Testimonies of the Testaments*], Heraklion: Πανεπιστημιακές Εκδόσεις Κρήτης [Crete University Press], 2000, pp. 66–67; Yannitsiotis, *The Social History of Piraeus, 1860–1910*, p. 64. With this in mind, the "landowners" found in the registers are classified in the agricultural sector only if they have no other occupational designation.

[18] This group includes only soldiers and sailors for whom no other occupation is recorded.

Table 7.1 (continued)

Economic sector	*Occupations*
Services	Bankers, brokers, estate agents
	Cooks, barbers, innkeepers, pawnbrokers
	Employees and servants in hotels, restaurants, cafes (e.g. waiters and waitresses)
	Prostitutes
Domestic services	–
	Housekeepers
	Domestic servants[19]
Commerce	Wholesale traders
	Suppliers
	Shopkeepers and traders (wine merchants, grain merchants, etc.)
	Clerks, shop assistants and shop servants (e.g. grocery shop assistants), retailers
	Street vendors
Transportation	Ship engineers, captains and navigators
	Seafarers, shipbuilders, ship and vehicle builders and craftsmen (for example, "marine ironmonger" and "tramway electrician")
	Seamen, cabin crew and workers in ships (not in the navy), ports and shipyards
Professions	Doctors, professors, lawyers, notaries, PhDs, archaeologists
	Nurses, pharmacists, teachers, educators, journalists, interpreters, artists, priests, monks
	–
	–
	–
No occupation and no indication of social status	Housewives, pupils and students, patients for whom it was declared that they were not working
Unknown occupation	Patients with no occupation recorded or without sufficient data to be assigned to a group

Source: Samples from the registers, Dromokaitio and Eginitio, 1905–1909 and 1916–1920 (see Table 1.1).

[19] Although in Greece those who described themselves as servants were often employees who lived in their employer's house, by the early twentieth century the term "servant" had largely come to be identified with female domestic service and was distinguished from the terms "employee" and "worker". Yannitsiotis, *The Social History of Piraeus, 1860–1910*, pp. 70, 72. For an overview of women's salaried domestic work in the first half

landlords,[20] is included in the first subgroup because the number of landowners without any other occupational designation was small; this can support the hypothesis that there was a difference, however small, between the few patients who were recorded as landowners and the many who were recorded as farmers. The occupations in the second subgroup of each sector also correspond to the middle classes but more to the lower ones, both rural and urban, including the petty bourgeoisie[21] and the strata between the petty bourgeoisie and the working class: since many of the artisans and tradesmen were not proprietors and employers, they were probably closer to the working class rather than to the bourgeoisie.[22] Finally, the third subgroup of each economic sector includes unskilled and landless occupations, such as unskilled labourers (in industry, agriculture and mining), unskilled employees (e.g. cleaners and messengers), domestic servants and hawkers. All these occupations, again with reservations, refer to the lower social classes.

It is obvious that the categorisation into economic sectors, and even more so the tripartite social classification, is rather fluid, not only because of the limitations of the sources, but also because of the already unclear boundaries between social classes. Nevertheless, it can facilitate the understanding of the social physiognomy of the patients. Starting with the economic sectors (Table 7.2) and excluding housewives, which is the largest category but does not provide information on socio-economic status, most of the patients were occupied in industry, commerce, the public and private sectors and the armed forces. Agriculture, livestock and fisheries were only in fourth place, although Greece remained a largely rural country at the beginning of the twentieth century.[23] Thus, the fact

of the twentieth century, see Pothiti Hantzaroula, *Σμιλεύοντας την Υποταγή. Οι Έμμισθες Οικιακές Εργάτριες στην Ελλάδα το Πρώτο Μισό του Εικοστού Αιώνα* [*Forging Submission. Paid Domestic Workers in Greece in the First Half of the Twentieth Century*], Athens: Παπαζήσης [Papazisis], 2012.

20 Yannitsiotis, *The Social History of Piraeus, 1860–1910*, p. 64.

21 On the petty bourgeoisie, see Raptis, "Middle classes and middle-class culture in Europe", p. 227.

22 Indeed, for some historians, the artisans, merchants and lower clerks belonged to the upper working class and not to the petty bourgeoisie. Thomas Martin Devine, *The Scottish Nation 1700–2000*, London: Penguin Books, 1999, p. 369.

23 In 1920, 57.5 % of the active population was employed in agriculture and farming. Leda Papastefanaki, "Μισθωτή εργασία" ["Paid work"], in Kostas Kostis and Socrates Petmezas (eds), *Η Ανάπτυξη της Ελληνικής Οικονομίας κατά τον 19ο αιώνα (1830–1914)*

Table 7.2 Occupations of patients admitted to the Eginitio and the Dromokaitio by economic sector

Economic sector	*% of patients admitted*	
	Eginitio	*Dromokaitio*
Agriculture, livestock, fisheries	10.1	7.0
Industry/manufacturing	12.3	8.9
Building	1.1	1.1
Mining	0.1	0.0
Public and private sector and armed forces	11.5	25.3
Services	2.2	1.4
Domestic services	1.0	0.4
Commerce	11.2	11
Transportation	3.4	2.7
Professions	6.8	6.7
Pupils, students	5.7	5.2
Housewives	33.1	22.7
No occupation[24]	0.6	0.5
Unknown occupation	0.9	7.1
Total	100	100

Source: Samples from the registers, Eginitio and Dromokaitio, 1905–1909 and 1916–1920 (see Table 1.1).

that the majority of patients had urban occupations suggests that the rural part of Greek society used psychiatric institutions to a lesser extent than the urban part, especially when we consider that both institutions admitted patients from all over Greece, not just from Athens, as we shall see in the next section.

Table 7.2 further shows that the two institutions did not differ much from each other in terms of the occupations of the patients admitted. The main difference is that the Dromokaitio had a higher proportion

[*The Development of the Greek Economy in the 19th century (1830–1914)*], Athens: Αλεξάνδρεια [Alexandria], 2006, pp. 253–91.

24 This category only includes patients for whom the registers recorded no previous occupation that would allow their classification. If we take into account the patients who had a previous occupation recorded but were not working at the time of admission, 0.7% and 1.4% of the patients admitted to the Eginitio and the Dromokaitio, respectively, had no occupation. These percentages were probably higher, as for some patients whose occupation was recorded in the registers, it was noted in their case histories noted that they had stopped working before admission.

of patients working in the private and public sectors and in the armed forces, which was largely due to the numerous men from the armed forces admitted to the institution during the period studied.

Gender differences in patients' occupations followed the gendered characteristics of work that were prevalent at the time. On the one hand, female patients were employed in typical female occupations: they were primarily in the professions, mainly teachers, in industry/manufacturing, largely seamstresses and finally in domestic services. On the other hand, most of the female patients admitted were recorded as not working. This is indicated in Table 7.2 by the category "housewives", which is all female and accounted for the largest proportion of female patients: 85% of women admitted to the Eginitio and 81% to the Dromokaitio. However, case histories suggest that women's occupations were under-recorded in the registers, as in the general population censuses (see Chapter 2). Some women who were listed as housewives in the registers were noted in the case histories as working, usually in agriculture,[25] textiles, like Eugenia P., who had been embroidering for eight hours a day since she was 15 years old,[26] or domestic service: for example, Penelope M., in addition to washing her family's clothes, also washed those of other families,[27] and Billio M., because she was poor, was forced not only to do all the household work alone, but also to take on outside work.[28]

The breakdown of each sector into the three subgroups also suggests common trends between the two institutions. Although the Dromokaitio was more similar to a public institution than the Eginitio in the early twentieth century (see Chapter 4), it seems that the two hospitals were not so strongly differentiated in terms of the socio-economic status of their patients. In both hospitals, most patients (42%–37%) belonged to the second subgroup of each economic sector (mainly craftsmen, artisans, low-ranking employees, shopkeepers and shop assistants). They were followed by patients with occupations in the first and third subgroups (10%–11% and 5%–8% of patients, respectively). In both institutions, as evidenced by occupations such as factory owners, engineers, bankers and high-ranking officials in the armed forces and the public sector, as well as

[25] Eg/9.2, pp. 301, 429.

[26] Eg/9.2, p. 405.

[27] Eg/9.2, p. 163.

[28] Eg/9.2, p. 143.

the high hospital fees, some patients may have been relatively well-off and belonging to the upper social classes. Still, they represented only a small proportion of the patients admitted. Besides, only a minority of patients were treated in the first—most expensive—class and numerous patients found it difficult to afford the cost of hospitalisation (see Chapter 4). At the same time, the low presence of patients with occupations from the third subgroup, which is associated with the lower social strata, was to be expected due to the limited possibilities of free hospitalisation (see Chapter 4). Overall, combining the findings from the occupations with those from the analysis of the economic aspects of hospitalisation, it can be concluded that the Eginitio and the Dromokaitio were neither luxury nor pauper hospitals, but were predominately used by the middle and lower middle classes.

This picture is supported by details of the patients' socio-economic status provided in the case histories. These reveal that there were a few wealthy patients in the institutions, particularly in the Eginitio, where, for example, a landowner was described as "not deprived of property".[29] More often, however, patients were noted to have economic problems, and those without adequate support networks, usually widows without children, abandoned women and orphans, were worse off.[30] Economic problems were linked to work-related difficulties, which, in turn, were inextricably linked to poor health, causing some to neglect their work and others to be unable to perform it.[31] Finally, case histories stress poverty and tedious work in the cases of working women, but they also record the way in which women responded to domestic tasks: an inability or unwillingness to perform their domestic "duties" was seen as a sign of illness.[32]

Place of Residence

Another provoking or exacerbating factor of insanity, according to nineteenth- and early twentieth-century psychiatric discourse, was the place of residence. The form and frequency of mental illness were thought

[29] Eg/9.2, p. 389.

[30] For example, Eg/9.2, p. 387.

[31] For example, Dr/Case Histories, 1917, p. 11; 1918, pp. 50, 67.

[32] For example, Dr/Case Histories, 1907, p. 17.

to be influenced, for example, by regional characteristics, such as whether an area was mountainous, lowland, or marshy, and by the climate.[33] Yet the most intense debates on the relationship between the environment and mental illness were intertwined with the issue of "civilisation" and in particular the influence of the type of community: rural, urban and industrial. Some doctors believed that the rate of insanity was higher in rural areas, but most considered the city to be more dangerous.[34] Aspects inextricably linked with urban life were commonly blamed for mental illness: unsanitary conditions and overcrowding, irregular and competitive lifestyle, intellectual professions, substance and sexual abuse, and immorality in general.[35] Doctors were particularly alarmed by life in busy harbours and the effects of shipping and trade, mixing with "lower-class" women, frequenting brothels and the deviation from traditional mores and ways of life.[36] Linking immorality, the weakening of traditions and excessive mental activity with madness, psychiatric theory deployed and reinforced fears of urban life and defended the ideal of traditional, quiet and temperate rural life.

If the psychiatric discourse related the place of residence to the prevalence and typology of mental illness, the remainder of the section uses this to better understand psychiatric hospitalisation, particularly the question of who was hospitalised, and the social position of the institutions. In this respect, it is of particular interest whether the patients came from urban or rural areas, not to examine whether mental illness was more prevalent in cities, but whether psychiatric hospitalisation was a more familiar,

[33] Esquirol, *Des Maladies Mentales*, v. 1, pp. 24–25; Katsaras, *Pathology of the Nerves and Psychiatry*, v. 2, p. 399.

[34] On this debate see John Charles Bucknill and Daniel Hack Tuke, *A Manual of Psychological Medicine: Containing the History, Nosology, Description, Statistics, Diagnosis, Pathology, and Treatment of Insanity: With an Appendix of Cases*, London: J. Churchill, 1858, pp. 251–52.

[35] Henry Maudsley, *The Physiology and Pathology of the Mind*, London: Routledge/ Thoemmes Press, 1993 (first edition in 1867), p. 204.

[36] Yeoryios Zilanakis, "Η προϊούσα καθολική παράλυσις. Στατιστική της δεκαπενταετίας 1908–1922" ["General paralysis of the insane. Statistic of the fifteen years 1908–1922"], *Ιατρική Πρόοδος* [*Medical Progress*], 30 (1925), pp. 204–7. Zilanakis was chief physician of the Mental Hospital of the National Hospitals of Constantinople and was writing about the specific harbour city. His paper was not about mental illness in general but about the general paralysis of the insane (see Chapter 8), whose syphilitic causation—established by the early twentieth century—associated it with the "immoral" lifestyles of cities and harbours.

possible and acceptable prospect for urban populations. In addition, I use the place of residence recorded in the registers as an indicator of the distance patients travelled to be admitted and, therefore, of the geographical range within which admission to the Eginitio and the Dromokaitio was a feasible option.

Before moving on to the analysis, it should be noted that there are certain problems with the way that patients' residence was recorded in the registers. Often the place of residence was not recorded, and sometimes it was contradicted by other information from the registers or the case histories. For these reasons, the analysis is limited to the cases in which residence is recorded and relatively certain. Also, the classification of residence as urban and rural is particularly problematic for patients for whom a wide area was recorded, for example Macedonia or Egypt. Furthermore, in many cases it cannot be affirmed whether that the exact place of residence of the patient was recorded, rather than the nearest town. A more serious problem is that in some cases, especially in the Dromokaitio registers, the place of origin is recorded under the column "place of residence".[37]

Despite these problems, analysis of the available data can provide some insight into rural or urban residence and the geographical distribution of the patients. On the one hand, a slight majority of patients with a known place of residence lived in cities of more than 5,000 inhabitants before admission. Therefore, in combination with the analysis of occupations, it can be concluded that the institutions were mainly used by the urban population. On the other hand, it is possible to draw conclusions about the distance patients travelled to be hospitalised by classifying the places of residence into four groups, from the nearest to the furthest from Athens.[38] Most of the patients admitted lived in nearby areas (groups A and B), but there were patients from all over Greece and from areas outside the Greek state (Table 7.3). The wide reach of the

[37] In 1834, Athens became the capital of the newly established Greek state (1828) and from then on it attracted numerous inhabitants from different places within and outside the country's borders. In the first part of the twentieth century, only 25% of Athenians were born in Athens. Bournova, *The Residents of Athens (1900–1960)*, p. 103. Therefore, many patients who resided in Athens had other places of origin.

[38] One problem with this classification is that it does not consider the transport networks between the place of residence and the hospitals, nor the topography of the routes, for example whether they crossed the sea or high mountains. However, these four zones give an overview of the distance travelled by patients in order to be hospitalised.

institutions was certainly related to the lack of a geographically developed network of mental health care services in early twentieth-century Greece (see Chapter 3). This extended function was more pronounced in the Dromokaitio, but this may have been due to the large number of army patients, who came from all over Greece, as well as to the problem of registers—more pronounced in the Dromokaitio—of recording the place of origin rather than the place of residence. The same problem is probably the reason why there were fewer patients from group A than from group B in the Dromokaitio. In any case, it seems that patients living far from the institutions constituted a noteworthy, albeit minor, component of admissions to both institutions.

The relationship between place of residence and hospitalisation can be better analysed with the help of the data from the Eginitio, since these are more reliable. There, patients who travelled further to reach the hospital (groups C and D) tended to be male, adult (26–39 years old) and unmarried. Thus, it could be argued that women, children, the elderly and married people were less likely to travel long distances in order to be hospitalised. In addition, patients admitted from distant areas tended to stay longer in hospital and were more likely to be transferred to a lower

Table 7.3 Place of residence of the patients admitted to the Eginitio and the Dromokaitio by distance from Athens

Group	*Eginitio*		*Dromokaitio*	
	Number of patients	*% of patients admitted*	*Number of patients*	*% of patients admitted*
A	563	40.8	279	29.6
B	417	30.2	366	38.8
C	224	16.2	220	23.3
D	176	12.8	79	8.4
Total of patients with known place of residence	1380	100.0	944	100.0

Source: Samples from the registers, Eginitio and Dromokaitio, 1905–1909 and 1916–1920 (see Table 1.1).

Group A: Attica. Group B: Central Greece, Thessaly, Peloponnese, Cyclades. Group C: Ionian Islands, Macedonia, Epirus, Thrace, Crete, East Aegean islands. Group D: all regions outside the borders of the Greek state after 1913: Asia Minor, Constantinople, Dodecanese, Cyprus, Egypt, Bulgaria, Romania, Russia and America

hospital class, in order to prolong their stay. Finally, patients from groups C and D were less likely to be readmitted.

It follows that the distance of home from the institutions affected their use. Long distances created conditions for long-term rather than repeated hospitalisation and discouraged the admission of women, children, the elderly, and married people. For these categories, who, as argued in the previous two chapters, were generally less likely to be admitted to mental hospitals, the long distance between home and institution made admission even more inconvenient, inappropriate and, ultimately, undesirable. An institution far from home was less welcoming and accessible, meant higher costs and difficulties in the movement of patients and visitors and, therefore, limited contact between patients and their relatives and friends. This, in turn, could mean a greater sense of alienation of patients from their private environment.

This chapter highlighted that occupation and work in general, as well as place of residence, played an important role in the way mental illness was understood and represented, being seen as causal factors or indicators of mental illness. Increased mental morbidity was attributed to urban life and certain occupations, as well as to the work of women beyond their accepted gender roles. The association of mental illness with urban life and work/occupation is interesting in itself, as it underlines once again the social and moral dimensions of psychiatric discourse, as well as the continuity and cross-fertilisation of medical and lay conceptions of mental illness.

However, as has been pointed out several times already, the data from the institutions are not so much an indication of who was ill as of who was hospitalised. The information on occupations and place of residence, together with the data analysed in Chapter 4, draw a relatively complete picture of the social composition of the population of the institutions. Although the patients came from all over Greece, a predominately rural country, most of them had occupations that were typical of the urban middle and lower middle classes. In terms of socio-economic status, there were few patients from the highest socio-economic strata and fewer from the lowest, although many patients faced financial problems and found it difficult to pay for their hospitalisation. The time of illness and hospitalisation was difficult for many patients and their families and, as the case histories show, the challenges patients faced in carrying out their tasks within and outside the home were often both a starting point and an important aspect of the economic and social deterioration that mental illness entailed.

Finally, the very character of the psychiatric institutions shaped their population and use. The absence of an extended psychiatric system in early twentieth-century Greece meant that the two Athenian hospitals had a wide geographical jurisdiction and can be described as national institutions. Although the long distance between home and the institution did not cancel out the possibility and potential of hospitalisation, it acted as a deterrent, especially for the admission of patients who were generally more difficult to be hospitalised, and contributed to longer hospital stays and the less readmissions. Furthermore, despite the national scope of the institutions, the analysis of both place of residence and occupation revealed a stronger presence of the urban population in admissions. This finding may support the argument that although psychiatric institutions also served rural populations, they were more commonly used by urban residents, and that rural communities were more reluctant to seek institutional mental health care.

CHAPTER 8

Diagnoses and Symptoms

This chapter focuses on the two main psychiatric categories through which doctors described mental illness: diagnoses and symptoms. These categories were constituted and deployed within specific social and scientific contexts and thus embodied and reflected social and scientific perceptions and values.[1] As such, they may say more about the theory

[1] Joan Busfield, "Class and gender in twentieth-century British psychiatry: Shell-shock and psychopathic disorder", in Jonathan Andrews and Anne Digby (eds), *Sex and Seclusion, Class and Custody: Perspectives on Gender and Class in the History of British and Irish Psychiatry*, Amsterdam: Rodopi, 2004, pp. 295–322; Jonathan Sadowsky, *Imperial Bedlam: Institutions of Madness in Colonial Southwest Nigeria*, Berkeley: University of California Press, 1999, pp. 69–76, 116–17; Catherine Coleborne, "'She does up her hair fantastically': The production of femininity in patient case books of the lunatic asylum in 1860s Victoria", in Jane Long (ed.), *Forging Identities: Bodies, Gender and Feminist History*, Nedlands: University of Western Australia Press, 1997, pp. 47–68; Jill Matthews, *Good and Mad Women: The Historical Construction of Femininity in Twentieth-Century Australia*, St. Leonards: Allen and Unwin, 1984. For a recent reappraisal of the process and challenges of diagnosis standardisation in nineteenth- and early twentieth-century Britain, see Kevin Matthew Jones, *Psychological Classification and Diagnosis in Asylum Statistics, 1800–1948. The British Table of the Forms of Insanity*, Cham: Palgrave Macmillan, 2025.

D. Kritsotaki, *A Social History of Mental Health Care in Greece, c. 1900–1920*, Mental Health in Historical Perspective,
https://doi.org/10.1007/978-3-032-05764-8_8

and practice of psychiatry than they do about the patients themselves.[2] Following this line of argument, my analysis does not treat diagnoses and symptoms as "objective" clinical data showing what the patients admitted to the institutions were suffering from.[3] However, I argue that they can contribute to an understanding of the physiognomy of the patients and the conditions that led to psychiatric hospitalisation, at least as seen from the doctors' perspective. To this end, the analysis is structured around the clarification of the psychiatric terms used in the Eginitio and the Dromokaitio in the early twentieth century. Each section of the chapter centres on a group of diagnoses, except for the last one, which synthesises the main symptoms recorded in case histories.

Diagnoses

At the beginning of the twentieth century, there were several systems for classifying mental illness, but none of them was internationally or even nationally accepted.[4] However, most classifications were based on a shared pool of symptoms that constituted a common psychiatric

[2] James Mills, "The mad and the past: Retrospective diagnosis, post-coloniality, discourse analysis and the asylum archive", *Journal of Medical Humanities*, 21 (2000), pp. 141–58.

[3] This—the retrospective diagnosis of patients based on the clinical descriptions found in hospital archives—has been attempted by a number of researchers. See, for example, Edward Renvoize and Allan Beveridge, "Mental illness and the late Victorians: A study of patients admitted to three asylums in York, 1880–1884", *Psychological Medicine*, 19 (1989), pp. 19–28; R. Parker, A. Dutta, R. Barnes and T. Fleet, "County of Lancaster Asylum, Rainhill: 100 years ago and now", *History of Psychiatry*, 4 (1993), pp. 95–105; Trevor Turner, "A diagnostic analysis of the casebooks of Ticehurst House Asylum, 1845–1890", Monograph Supplement 21, Cambridge: University Press, 1992; Allan Beveridge, "Madness in Victorian Edinburgh: A study of patients admitted to the Royal Edinburgh Asylum under Thomas Clouston 1873–1908", part 2, *History of Psychiatry*, 6 (1995), pp. 431–51; G. A. Doody, Allan Beveridge and E. C. Johnstone, "Poor and mad: A study of patients admitted to the Fife and Kinross District Asylum between 1874 and 1899", *Psychological Medicine*,26 (1996), pp. 887–98.

[4] As the French psychiatrist Édouard Toulouse (1865–1947) famously remarked in 1895, when psychiatrists could do nothing else, they made up a new classification. Patricia Prestwich, "Drinkers, drunkards, and degenerates: The alcoholic population of a Parisian asylum, 1867–1914", *Histoire Sociale/Social History*, 27 (1994), pp. 321–35. In 1905, the British psychiatrist Maurice Craig (1866–1935) counted at least ten major psychiatric classifications since the time of Philippe Pinel (1745–1826). R.M. Ion and M.D. Beer, "The British reaction to dementia praecox 1893–1913", part 1, *History of Psychiatry*, 13 (2002), pp. 285–304.

code, descriptive psychopathology, that emerged between the early nineteenth and twentieth centuries.[5] Leaving aside the subtle distinctions and different theories about the causes, course, outcome and pathological anatomy of each type of mental illness, this chapter analyses the psychiatric diagnoses recorded in the registers of the Eginitio and the Dromokaitio as headings of descriptive psychopathology, namely as symptom codes referring to the main disorders that doctors recognised in their patients.

To better manage the various diagnoses recorded in the registers, I have grouped them into broader categories based on their main symptoms (Tables 8.1 and 8.2). For example, the category of "melancholy" includes not only the simple diagnosis of "melancholy" but also various forms of melancholy and diagnoses which, according to the psychiatric theory at the time, had melancholy as their main symptom, such as "anxious melancholy" and "senile melancholy". This type of grouping serves the purpose of my analysis, which is not to examine the psychiatric diagnoses as such, but to identify through them the main characteristics that, according to the doctors, constituted the patients' condition. Subsequently, to offer a clearer picture of the meaning of the diagnoses, I turn to their descriptions by psychiatrists of the time, using internationally influential and Greek texts, especially those written by the directors of the Eginitio and the Dromokaitio.

Mania, Melancholy and Their Mixed, Alternating and Periodic Forms

Mania and melancholy have been among the oldest and broadest terms in psychiatry but were narrowed down to specific diseases of emotion and action in the mid-nineteenth century.[6] In the beginning of the next century, mania featured extreme cheerfulness and self-confidence, psychomotor agitation, irritability, sexual excitability, lack of awareness

[5] German Berrios, *The History of Mental Symptoms: Descriptive Psychopathology since the Nineteenth Century*, Cambridge: Cambridge University Press, 1995, pp. 15–17.

[6] Berrios, *The History of Mental Symptoms*, pp. 298–300. In general, this chapter explores the psychiatric meanings of the diagnoses, which began to take shape on the basis on observations made in the asylums in the context of the new medical speciality for the mentally ill that emerged at the end of the eighteenth century.

Table 8.1 Diagnostic categories, Eginitio

Diagnostic category	*Male patients*		*Female patients*		*All patients*	
	N	*%*	*N*	*%*	*N*	*%*
Melancholy	50	5.7	64	11.5	114	8.0
Mania	7	0.8	11	2.0	18	1.3
Intermittent insanity*	28	3.2	45	8.1	73	5.1
Psychomotor agitation	4	0.5	0	0.0	4	0.3
Delusional states	55	6.3	45	8.1	100	7.0
Degeneration/degenerative insanities	13	1.5	11	2.0	24	1.7
Mental weakness/deficiency/retardation	2	0.2	1	0.2	3	0.2
Dementia praecox	120	13.7	72	13.0	192	13.4
Catatonia	8	0.9	3	0.5	11	0.8
Pseudo dementia praecox	9	1.0	9	1.6	18	1.3
Dementia	16	1.8	13	2.3	29	2
Senile insanity	3	0.3	0	0.0	3	0.2
Presenile insanity	0	0.0	1	0.2	1	0.1
Mental confusion	36	4.1	34	6.1	70	4.9
Insanity	0	0.0	1	0.2	1	0.1
General paralysis of the insane (GPI)	125	14.3	9	1.6	134	9.4
Hysteria (alone or combined with other diagnoses)	10	1.1	79	14.2	89	6.2
Epilepsy/epileptoid attack	18	2.1	18	3.2	36	2.5
Neurasthenia	23	2.6	8	1.4	31	2.2
Hypochondria	1	0.1	6	1.1	7	0.5
Neurological/bodily diagnoses**	295	33.7	119	21.4	414	29.0
Alcoholic diagnoses	33	3.8	3	0.5	36	2.5
Toxic diagnoses	18	2.1	3	0.5	21	1.5
No diagnosis/unidentified diagnosis	1	0.1	0	0.0	1	0.1
Total	875	100.0	555	100.0	1430	100.0

*The term "insanity" translates the Greek word "phrenitis" and "phrenopathy", meaning illness of the brain. In the Eginitio and in the Dromokaitio (from 1918) the terms "psychosis" and "psychopathy" were also used. See also Chapter 1, "Terminology"
**These diagnoses had mainly physical symptoms, such as pain, paralysis and fever, including,hemiplegia, meningitis and multiple sclerosis. See below, section "Nervous illnesses"
Source: Sample from the registers, Eginitio, 1905–1909 and 1916–1920 (see Table 1.1).

of illness, babbling and incoherence, and exacerbation of evil tendencies such as slander.[7] Melancholy had the opposite symptoms: withdrawal,

[7] Michael Katsaras, *Παθολογία των Νεύρων και Ψυχιατρική. Νόσοι του Εγκεφάλου, Νόσοι της Παρεγκεφαλίδος, Φρενίτιδες Νόσοι* [*Pathology of the Nerves and Psychiatry. Diseases of*

Table 8.2 Diagnostic categories, Dromokaitio

Diagnostic category	*Male patients*		*Female patients*		*All patients*	
	N	*%*	*N*	*%*	*N*	*%*
Melancholy	30	4.3	30	11.2	60	6.2
Mania	23	3.3	17	6.3	40	4.2
Intermittent insanity	63	9.1	44	16.4	107	11.1
Psychomotor agitation	15	2.2	9	3.3	24	2.5
Delusional states	74	10.7	39	14.5	113	11.8
Degeneration/degenerative insanities	86	12.4	23	8.6	109	11.3
Moral insanity	1	0.1	0	0.0	1	0.1
Mental weakness/deficiency/retardation	24	3.5	6	2.2	30	3.1
Dementia praecox	99	14.3	50	18.6	149	15.5
Catatonia/catatonic cataplexy	1	0.1	1	0.4	2	0.2
Dementia	24	3.5	15	5.6	39	4.1
Presbyophrenia	0	0.0	1	0.4	1	0.1
Mental confusion	23	3.3	5	1.9	28	2.9
General paralysis of the insane (GPI)	133	19.2	12	4.5	145	15.1
Epilepsy/epileptoid attack	13	1.9	11	4.1	24	2.5
Hysterical psychopathy	0	0.0	1	0.4	1	0.1
Neurasthenic psychosis	1	0.1	0	0.0	1	0.1
Neurological/bodily diagnoses	17	2.5	1	0.4	18	1.9
Alcoholic diagnoses	50	7.2	0	0.0	50	5.2
Toxic diagnoses	7	1.0	2	0.7	9	0.9
No diagnosis/unidentified diagnosis	8	1.2	2	0.7	10	1.0
Total	692	100.0	269	100.0	961	100.0

Source: Sample from the registers, Dromokaitio, 1905–1909 and 1916–1920 (see Table 1.1).

morosity and worry, lack of energy and courage, mental and psychomotor slowness, passivity, but also delusions and hallucinations.[8] According to some doctors, another symptom of melancholy, which signalled the most

the Brain, Diseases of the Cerebellum, Phrenic Diseases], v. 2, Athens: Εστία [Estia], 1898, pp. 439–59; Emil Kraepelin, *Cent Ans de Psychiatrie, Suivi de La Folie Maniaco-Depressive* (1917), transl. Marc Geraud, Bordeaux: Mollat, 1997, pp. 199–236; Christopher Baethge, Ira Glovinsky and Ross Baldessarini, "Manic-depressive illness in children: An early twentieth-century view by Theodor Ziehen (1862–1950)", *History of Psychiatry*, 15 (2004), pp. 201–26.

[8] Berrios, *The History of Mental Symptoms*, pp. 294–300; Kraepelin, *Cent Ans de Psychiatrie*, pp. 199–236; Katsaras, *Pathology of the Nerves and Psychiatry*, v. 2, pp. 374–88.

serious form of the illness, was cataplexy. This was characterised by stillness, speechlessness, inhibition of emotions and resistance to or passive acceptance of any demand from the environment.[9]

In the first half of the nineteenth century, many doctors noted that these two opposite clinical conditions could occur in the same person through the combination or alternation of manic and melancholic symptoms.[10] At the end of the nineteenth century, the French psychiatrist Valentin Magnan (1835–1916) created the diagnosis of intermittent or periodic insanity, characterised by periodic episodes of mania and/or melancholy,[11] while the German psychiatrist Emil Kraepelin (1856–1926) described manic-depressive insanity, which included both single and mixed forms of mania and melancholy.[12]

The attributes coded by the diagnoses of melancholy, mania and their mixed and periodic forms (grouped together for the purposes of the chapter as "intermittent insanity") were among the most frequently identified in the institutions, although mania was diagnosed less frequently than melancholy (Tables 8.1 and 8.2). In addition, some patients were diagnosed with "agitation", and it is therefore possible that their symptoms were similar to those of mania. Of the mixed and periodic diagnoses, "intermittent insanity" was the most common, while "manic-depression" was very rare and was only noted as a form of "intermittent insanity".

As shown in Tables 8.1 and 8.2, the diagnostic groups of melancholy, mania and intermittent insanity were populated by more women than men, a trend underlined by psychiatrists of the time and attributed to the "nature" of women, mainly their nervous system and the organic and emotional processes during puberty, pregnancy, postpartum and

[9] Katsaras, *Pathology of the Nerves and Psychiatry*, v. 2, pp. 427–37; Mercier, *A Text-book of Insanity*, pp. 191–97.

[10] Edward Shorter, *A Historical Dictionary of Psychiatry*, Oxford: Oxford University Press, 2005, pp. 165–66; Christopher Baethge, Paola Samvatore and Ross Baldessarini, "Ewald Hecker, cyclothymia, a circular mood disorder", *History of Psychiatry*, 14 (2003), pp. 377–90.

[11] Katsaras, *Pathology of the Nerves and Psychiatry*, v. 2, pp. 460–67; Michael Yanniris, *Αι Φρενοπάθειαι εν Ελλάδι: Μελέτη επί Στατιστικών και Κλινικών Παρατηρήσεων* [*The Phrenopathies in Greece: A Study on Statistical and Clinical Observations*], Syros: s.n., 1898, p. 27.

[12] Shorter, *A Historical Dictionary of Psychiatry*, pp. 166–67; K. Trede et al., "Manic depressive illness: Evolution in Kraepelin's textbook, 1883–1926", *Harvard Review of Psychiatry*, 13 (2005), pp. 155–78.

menopause.[13] Moreover, it was claimed that women suffered from melancholy more than men because "scarcity of necessities and domestic sorrows are more often encountered and felt in women".[14] Women were also considered more prone to intermittent and periodic forms of melancholy and mania because of their alleged greater emotional irritability and periodic nature, as evidenced by menstruation. The widespread psychiatric belief that women were more emotionally vulnerable than men was in line with perceptions of the different characteristics of the sexes and, in the early twentieth century, was backed by the observations of almost a century of clinical psychiatry. However, the assumption of emotional sensitivity as a female trait had played a role in the production of these clinical observations and continued to influence diagnostic practice, contributing to women being diagnosed with affective disorders more frequently than men. After all, many of the symptoms of these diagnoses, such as irritability, talkativeness, difficulty in concentrating, crying, petty theft and fastidiousness in dressing, were closely tied to femininity.[15] Nevertheless, the association of affective disorders with gender was not clear cut: some symptoms, especially of mania and intermittent insanity, were more bound up with masculinity, such as violent, destructive and dangerous acts (see Chapter 5) and these diagnoses were certainly given to men as well. At the same time, women in acute attacks of mania and intermittent insanity were also described as extremely noisy, violent, volatile, impulsive, and as indecent and obscene in speech and behaviour.[16]

The most clearly gendered manifestations of affective disorders related to sexual morality. In female patients, symptoms were often associated

[13] Henry Maudsley, *The Physiology and Pathology of the Mind*, London: Routledge/Thoemmes Press, 1993 (first edition in 1867), p. 212; Havelock Ellis, "Sex, influence of, in insanity", in Daniel Hack Tuke (ed.), *A Dictionary of Psychological Medicine: Giving the Definition, Etymology and Synonyms of the Terms used in Medical Psychology with the Symptoms, Treatment, and Pathology of Insanity and the Law of Lunacy in Great Britain and Ireland*, v. 2, London: J. & A. Churchill, 1892, pp. 1152–56; Hilary Marland, "Disappointment and desolation: Women, doctors and interpretations of puerperal insanity in the nineteenth century", *History of Psychiatry*, 14 (2003), pp. 303–20.

[14] Katsaras, *Pathology of the Nerves and Psychiatry*, v. 2, p. 398.

[15] Elizabeth Lunbeck, *The Psychiatric Persuasion: Knowledge, Gender, and Power in Modern America*, Princeton, N.J.: Princeton University Press, 1994, pp. 147–51.

[16] Ornella Moscucci, *The Science of Woman: Gynaecology and Gender in England, 1800–1929*, Cambridge: Cambridge University Press, 1990, p. 106.

with uncontrolled femininity and erotic behaviour, which was seen as different from that of men. Men manifested their "stimulation of the sexual instinct" in a direct way, for example by going to brothels or talking directly about sexual matters. Women, on the other hand, usually expressed their sexual arousal indirectly: they dressed and adorned themselves eccentrically, used "ambiguous" words, talked about marriages and love scandals, plotted erotic schemes and falsely accused those around them of obscene propositions, seduction or rape.[17] The different expressions of pathological and uncontrolled eroticism and sexuality in men and women exemplify the relationship between psychiatric and social representations, as the content of the disorder depended on the perceived normal expression of sexuality in each sex, the boundaries of which were clearly narrower for women. Moreover, as we shall see in the neurological diagnoses, sexual disorders in women were more often described at the behavioural level as overt, uninhibited and inappropriate eroticism, whereas in men the emphasis was on the physical aspect of organic dysfunctions.

Delusional States

According to one psychiatric definition, delusions were beliefs "in something that would be incredible to people of the same class, education, or race as the person who expresses it, the belief persisting in spite of proof to the contrary, this resulting from diseased working of the brain convolutions".[18] The social situation of the patients was considered to play a role in the content of the delusion. For example, villagers usually expressed religious delusions, such as being saints, while educated people expressed scientific delusions, such as being magnetised or electrified.[19] In other words, psychiatric theory of the time connected delusions to the social and cultural context of those who expressed them, although it also asserted that some delusions were witnessed in every era and society.

Delusional thoughts were noted as a symptom of many disorders, but from the early nineteenth century delusion began to be seen as a separate

[17] Katsaras, *Pathology of the Nerves and Psychiatry*, v. 2, pp. 451–52.

[18] Thomas Clouston, *Clinical Lectures on Mental Diseases*, 3rd edition, London: J. & A. Churchill, 1892, p. 244.

[19] Yanniris, *The Phrenopathies in Greece*, pp. 33–34; Katsaras, *Pathology of the Nerves and Psychiatry*, v. 2, p. 487.

mental disorder (Paranoia in German, délire in French),[20] characterised by distortion of the intellect without disturbances of the emotions (such as irritability and fear) or of the will (such as a decrease in initiative and self-control).[21] Although delusion as a distinct disease had additional symptoms, such as psychomotor agitation, talkativeness, incoherence and hallucinations, the hallmark symptom was delusional ideas.[22] In terms of their theme, various forms of delusions were distinguished in the late nineteenth and early twentieth centuries: delusions of melancholy, fear, jealousy, persecution and grandeur (patients believing they were wealthy, powerful and/or endowed with outstanding abilities).[23] Delusions were thought to take different forms between the sexes. Men were described as more stable in their delusions, which were more often of grandeur, while women were noted to manifest more often erotic and religious delusions.[24]

The diagnosis of delusion was quite frequent in the Dromokaitio and the Eginitio, affecting men and women practically equally (Tables 8.1 and 8.2). In addition, delusional ideas were noted as a symptom in other diagnoses, such as melancholia, mania and general paralysis of the insane. It is therefore clear that the specific symptom was prominent in the psychiatric assessment and the profile of the patients admitted to the hospitals. The types of delusional diagnoses found in the two institutions were influenced by French psychiatric theory, as they included the "chronic systematised" delusion and the "immediate" delusion. The former had

[20] Shorter, *A Historical Dictionary of Psychiatry*, pp. 206–7, 242; Ian Dowbiggin, "Delusional diagnosis? The history of paranoia as a disease concept in the modern era", *History of Psychiatry*, 11 (2000), pp. 37–69.

[21] Katsaras, *Pathology of the Nerves and Psychiatry*, v. 2, p. 484; Ian Dowbiggin, "Back to the future: Valentin Magnan, French psychiatry, and the classification of mental diseases, 1885–1925", *Social History of Medicine*, 9 (1996), pp. 383–408.

[22] Katsaras, *Pathology of the Nerves and Psychiatry*, v.2, pp. 617–23.

[23] Pierre Deniker and Jean-Pierre Olie, *Fou, Moi? La Psychiatrie Hier et Aujourd'hui*, Paris: O. Jacob, 1998, p. 93; Clouston, *Clinical Lectures*, pp. 251–79; Valentin Magnan and Paul Légrain, "Delusion in degenerates. The immediate delusions", in Francois-Regis Cousin, Jean Garrabé and Denis Morozou (eds), *Anthology of French Language Psychiatric Texts*, transl. John Crisp, Le Plessis-Robinson: Institut Synthélabo pour le Progrès de la Connaissance, 1999, pp. 267–83.

[24] Jean-Étienne Dominique Esquirol, *Des Maladies Mentales: Considérées sous les Rapports Médical, Hygiénique et Médico-Légal* (1838), v. 1, Paris: J.-B. Baillière, 1858, p. 38.

been described in the 1880s and 1890s by Valentin Magnan and Paul Serieux (1864–1947) and was characterised by delusions of persecution, progressing to delusions of grandeur and culminating in dementia.[25] The "immediate" delusion was described by Magnan in collaboration with Paul Légrain (1860–1939).[26] It was a sudden, polymorphous and incoherent delusion, considered to be an exclusive, and the most common, disorder of the "degenerates".[27] In Greek, it was translated as "sudden delusion of the degenerates" or "polymorphous" or "multiple" delusion. Other delusional diagnoses recorded in the Eginitio and the Dromokaitio registers were the acute, chronic, melancholic, hypochondriacal and hallucinatory delusion, the delusions of self-blame, persecution and grandeur; and paranoia, more rarely and only in the Dromokaitio.

Degeneration and Degenerative Insanities

Degeneration, a term used in natural history since the eighteenth century, became an analytical tool and a fundamental concern of Western political thought, philosophy and literature in the second half of the nineteenth century.[28] The concept was introduced into psychiatry by Bénédict-Augustin Morel (1809–1873), medical director of the St.-Yon asylum in France. For Morel, degeneration was the morbid deviation from an ideal primitive type of human being. This deviation, if not arrested by intermarriage with healthy individuals, was transmitted hereditarily, intensifying and often mutating from generation to generation.[29] In 1857, Morel described the typical progression in the generations of a degenerate family from a "nervous disposition" (tendency to irritability, violence and

[25] Yanniris, *The Phrenopathies in Greece*, pp. 31–35; Katsaras, *Pathology of the Nerves and Psychiatry*, v. 2, pp. 507–10.

[26] Valenin Magnan, *Leçons Cliniques sur les Maladies Mentales*, Paris: Publications de Progrès Medical, 1891; Valenin Magnan and Paul Légrain, *Les Dégénérés. État Mentale et Syndromes Episodiques*, Paris: Rueff, 1895.

[27] Magnan and Légrain, "Delusion in degenerates". The concept of degeneration is explored in the next section.

[28] On the many meanings of degeneration not just in medicine but also in politics, social research and the arts, see Daniel Pick, *Faces of Degeneration: A European Disorder, c.1848–c.1918*, Cambridge: Cambridge University Press, 1989.

[29] Patrice Pinell, "Dégénérescence", in Dominique Lecourt et al. (eds), *Dictionnaire de la Pensée Médicale*, Paris: Presses Universitaires de France, 2004, pp. 310–12.

exaggerations of character) to brain disorders and neuroses (hypochondria, hysteria, epilepsy), eccentricity and dangerousness, and, finally, intense and incurable physical, moral and mental degeneration, resulting to mental retardation, deafness and dumbness, insanity and, eventually, sterility, so that the family disappeared.[30] Behind this degenerative process were identified many physical and moral factors, such as unhealthy and unsuitable housing, diet and work; the use of alcohol, cannabis, opium and tobacco; famines and epidemics; even the composition of the soil and air.[31]

After Morel, the concept of degeneration became widespread and was further enriched. It was argued that degenerates were unbalanced, namely that the centres of their nervous system were unevenly developed. A detailed mapping of the "stigmata" of degeneration was elaborated, including both mental traits, such as mental retardation, obsessive ideas and impulses (actions performed consciously, but beyond the control of the degenerates), and physical characteristics, for example atrophies, skull deformities and dental abnormalities. Furthermore, in the course of the nineteenth century, degeneration lost the religious tone of Morel, for whom the primitive ideal man was Adam, and the beginning of degeneration was the fall from Eden, and incorporated Darwinian concepts, such as natural selection, and the struggle for survival.[32]

Degeneration theory was widespread in European psychiatry in the second half of the nineteenth and early twentieth centuries. Its success was largely due to its hereditary, biological and neurological reasoning, which, although unproven, fitted into the dominant biological psychiatry of the time and enhanced the scientific image and prestige of the psychiatric profession. Degeneration provided a unifying scheme that satisfied psychiatry's need for a robust model based on the cause of mental illness and helped to tidy up chaotic psychiatric theory. Moreover, at a time when the optimism of the early asylum doctors was waning due to the limited success of treatment and the ever-increasing asylum populations,

[30] Shorter, *A Historical Dictionary of Psychiatry*, p. 241; Pinell, "Dégénérescence", pp. 310–12.

[31] Bénédict-Augustin Morel, *Traité des Dégénéréscences Physiques, Intellectuelles et Morales de l'Éspece Humaine et des Causes qui produisent ces Variétés Maladives*, v. 1, Paris: J.-B. Baillière, 1857, pp. 47–65.

[32] Pinell, "Dégénérescence", p. 311.

the concept of insanity as a degenerative, i.e. hereditary and incurable, condition could justify the failures of psychiatry and the asylum.[33]

The concept of degeneration was widely used in Greek psychiatric theory and practice under the influence of French psychiatry. In the late nineteenth century, the textbooks of Michael Katsaras and Michael Yanniris (both published in 1898) treated degeneration as a distinct category and referred to a number of "degenerative hereditary insanities", including mental retardation (see next section), various cerebral and physical abnormalities, degenerative delusion (see previous section), as well as degenerative melancholia, mania, cataplexy and mental confusion, which were seen as variants of their non-degenerative counterparts. Finally, Yanniris and Katsaras classified under degeneration the "episodic syndromes" in which the mental stigmata of degeneration manifested themselves: lack of volition, which was expressed as the patient's inability to perform even simple activities, such as dressing or eating; obsessions, for example agoraphobia; impulses, such as dipsomania (the urge to drink intoxicating drinks), pyromania and the urge to commit homicide or suicide; and "sexual perversions" and abnormalities, such as masturbation, homosexuality, "lesbian eros", paedophilia, fetishism and "platonic eros". All these conditions were characterised by the subject's inability to control themselves and culminated in moral paranoia or moral insanity, in which the intellectual faculties were maintained in good condition, but the moral emotions were deficient or perverted. The morally insane were selfish, spiteful and quarrelsome; they did not love or respect their family and friends; they committed immoral and criminal acts (gambling, drinking, lying, slandering, and scheming, committing theft and murder); and they had no sense of the anti-social nature of their actions and no ability to correct themselves and be morally educated.[34]

In the Eginitio and the Dromokaitio, the most frequent degenerative diagnoses were degenerative insanities and obsessions, and, to a lesser extent, more general degenerative stigmata such as "imbalance", "eroticism", "perversion of the moral sense" and "moral insensitivity".

[33] Dowbiggin, "Back to the future", pp. 384, 389, 393.

[34] Katsaras, *Pathology of the Nerves and Psychiatry*, v. 2, pp. 523–92; Yanniris, *The Phrenopathies in Greece*, pp. 56–79. Moral insanity was first described in 1835 and consisted of moral perversions without mental disturbances, hallucinations or delusions. James Prichard, *A Treatise on Insanity and Other Disorders Affecting the Mind*, London: Sherwood, Gilbert, and Piper, 1835, p. 6.

In addition, there were patients who were considered "degenerate" but who received other diagnoses (mental retardation, melancholia, mania, delusion, dementia and alcoholic diagnoses). Taking them into account, the total percentage of "degenerates" reached 5.5% in the Eginitio and 16.1% in the Dromokaitio, with similar percentages for women and men. Therefore, although degeneration was used more in the Dromokaitio, the diagnostic practice in both hospitals confirms the prominence of degeneration in Greek psychiatry.

Mental Weakness/Deficiency/Retardation

In the late nineteenth and early twentieth century, mental weakness, deficiency or retardation was usually distinguished from insanity, and it was argued that sufferers should be treated in special institutions rather than in psychiatric asylums.[35] However, various forms of mental deficiency were included in the field of psychiatry: they were analysed in general psychiatric textbooks, and from the 1860s they became the subject of psychiatric specialisation in some countries, such as Britain and France.[36] Much of the psychiatric discourse on mental deficiency focused on children, because it was commonly detected at an early age. Indeed, the manifestations of mental deficiency were perceived by many doctors as the most common, or even the only, mental problems of childhood.[37] Besides, the mentally defective were often perceived and portrayed as children, regardless of their age.[38]

[35] William Ireland, *The Mental Affections of Children: Idiocy, Imbecility and Insanity*, 2nd edition, London: J. & A. Churchill, 1900 (first edition in 1898), pp. 419–20.

[36] David Wright, *Mental Disability in Victorian England: The Earlswood Asylum, 1847–1901*, Oxford: Clarendon Press, 2001, p. 169.

[37] Maudsley, *The Physiology and Pathology of the Mind*, p. 212; Ireland, *The Mental Affections of Children*, pp. 281–82.

[38] Mark Jackson, "'Grown-up children': Understandings of health and mental deficiency in Edwardian England", in Marijke Gijswijt-Hofstra and Hilary Marland (eds), *Cultures of Child Health in Britain and the Netherlands in the Twentieth Century*, Amsterdam and Atlanta: Rodopi, 2003, pp. 141–60. More generally, mental illness was widely associated with childhood. See, for example, Michel Foucault, *Madness and Civilization. A History of Insanity in the Age of Reason*, transl. Richard Howard, New York: Vintage, 1973, pp. 253–54; Jessie Hewitt, *Institutionalizing Gender. Madness, the Family, and Psychiatric Power in Nineteenth-Century France*, Ithaca: Cornell University Press, 2020, p. 74.

Mental retardation was classified into three degrees of severity: "idiocy", "stupidity", and "feeblemindedness", which was the least severe, being placed at the limit of normal intellectual development.[39] In all cases, organic problems were taken into account, such as cranial and physical deformities, as well as mental, emotional and moral characteristics.[40] Jean-Étienne Dominique Esquirol (1772–1840) was the first to define idiocy as the incurable and congenital absence of cognition, as opposed to dementia, which he regarded as a curable impairment of cognition, observed from puberty onwards.[41] Although idiots sometimes had certain talents, such as a strong memory,[42] they could not acquire the knowledge and skills of people in the same circumstances and had difficulty communicating.[43] In addition, they were characterised by a lack or perversion of their sensory faculties,[44] emotions[45] and instincts.[46] The mentally defective were often described as disobedient, arrogant, gluttonous, short-tempered and prone to crime, lying, alcohol abuse and sexual perversions, such as unrestrained sexual desire, masturbation, unnatural sexual acts, and exhibitionism.[47] The moral concept was

[39] Katsaras, *Pathology of the Nerves and Psychiatry*, v. 2, pp. 531–34.

[40] John Langdon Down, *On Some of the Mental Affections of Childhood and Youth*, London: J. & A. Churchill, 1887, pp. 23–24, 69; Clouston, *Clinical Lectures*, pp. 285–300.

[41] Jaqueline Gateaux-Mennecier, *Bourneville et l'Enfance Aliénée*, Paris: Centurion, 1989, pp. 78–80.

[42] Clouston, *Clinical Lectures*, pp. 285–300.

[43] Gateaux-Mennecier, *Bourneville et l'Enfance Aliénée*, pp. 58–61.

[44] For example, they could not feel cold or heat. Ibid., p. 102.

[45] For example, they were unable to love those who cared for them. Maurice Craig, *Psychological Medicine. A Manual of Mental Diseases for Practitioners and Students*, 3rd edition, London: J. Churchill, 1917, p. 347; Emmanuel Régis, *Précis de Psychiatrie*, 5th edition, Paris: Octave Doin et fils, 1914, pp. 529–35; Katsaras, *Pathology of the Nerves and Psychiatry*, v. 2, pp. 531–34.

[46] For example, they had an excessive or lacking appetite for food and intense sexual drives. Fletcher Beach, *The Treatment and Education of Mentally Feeble Children*, London: J. & A. Churchill, 1895, pp. 6–10.

[47] Gateaux-Mennecier, *Bourneville et l' Enfance Aliénée*, pp. 72, 147; Clouston, *Clinical Lectures*, pp. 285–300; Régis, *Précis de Psychiatrie*, pp. 527–35; Katsaras, *Pathology of the Nerves and Psychiatry*, v. 2, pp. 533–34; Craig, *Psychological Medicine*, pp. 347, 365.

particularly absent, defective or perverted in "moral idiocy", which was diagnosed, for example, in habitual criminals and children prone to vice.[48]

Evidently, the psychiatric discourse directly associated mental deficiency with social and moral problems, such as lack of judgement and self-control, intense sexuality, inability to work systematically, poverty, immorality, and crime.[49] Moreover, mental deficiency was perceived not only as a hereditary condition but also as an important component and form of degeneration. It is not surprising, then, that it fuelled fears and inspired attempts to control the mentally defective. These anxieties and endeavours intensified in several countries from the late nineteenth century, when concerns about the physical and mental robustness of the nation and race were at their peak and were shared by many psychiatrists.[50] Within this context, mental deficiency was included in eugenic research and interventions.[51] The eugenicists, among whom were many psychiatrists and specialists in mental deficiency, proposed, along with other things, the isolation of the mentally disabled in special institutions and the control of their fertility through premarital certification or sterilisation.[52] The idea of segregating the mentally defective in special institutions was in line with the reluctance of asylum authorities to accept such patients, who were considered a "burden" because of their

[48] Ireland, *The Mental Affections of Children*, p. 287; Henry Maudsley, "Body and mind: An inquiry into their connection and mutual influence, specially in reference to mental disorders" (1873), in Vieda Skultans, *Madness and Morals: Ideas on Insanity in the Nineteenth Century*, London: Routledge & Kegan Paul, 1975, pp. 246–52; Craig, *Psychological Medicine*, p. 364.

[49] Wright, *Mental Disability in Victorian England*, p. 189; Jan Walmsley, "Mental defectives in 1900: People with learning disabilities in 2000. What changed?: A response to Matt Egan's paper", in *Scottish Health History: International Contexts, Contemporary Perspectives*, Glasgow: Centre for the History of Medicine, University of Glasgow, 2003.

[50] Jackson, "'Grown-up children'".

[51] Walmsley, "Mental defectives in 1900"; Daniel Kevles, *In the Name of Eugenics: Genetics and the Uses of Human Heredity*, New York: Knopf, 1985, p. 77. In the early twentieth century, eugenics was a scientific and political ideology and movement with organisations in many countries, aiming to improve the human species through biological solutions of a wide range of problems, from tuberculosis and epilepsy to alcoholism, poverty and crime. See Kevles, *In the Name of Eugenics*; Pauline Mazumdar, *Eugenics, Human Genetics, and Human Failings: The Eugenics Society, its Sources and its Critics in Britain*, London: Routledge, 1992.

[52] Kevles, *In the Name of Eugenics*, ch. 6.

usually long hospital stays. In a number of countries, special institutions were founded and other of the above measures were implemented. On the contrary, in Greece no such action was taken, although eugenic views were supported, particularly in the interwar period, when eugenic proposals, such as the introduction of premarital certification, were widely debated.[53]

In the Eginitio and the Dromokaitio, there were few patients diagnosed with forms of mental deficiency and they were hospitalised for long periods of time, some for over ten years or until they died. This explains why in the Eginitio, which provided mainly short-term care, there were even fewer patients with mental deficiency. Moreover, in both institutions, most of these patients were diagnosed with mild degrees of mental deficiency but were recorded to have severe mental and nervous symptoms, such as delusions, melancholy, agitation and epileptic seizures. It was probably these symptoms, which made their condition more like insanity and epilepsy, that allowed their admission to the Eginitio and the Dromokaitio. In other words, the situation in the two hospitals reflected the psychiatric theory of the time, which called for the separation of the mentally defective from the insane, and the reluctance to admit patients to mental hospitals on the grounds of mental deficiency alone.

Finally, although diagnoses of mental deficiency were given to men and women of all ages in both institutions, children in the Dromokaitio were almost exclusively diagnosed with such conditions, which confirms the strong association between mental deficiency and childhood. The Eginitio was a different case: due to its neurological specialisation, most of the children there were diagnosed with nervous conditions, such as hysteria, chorea, epilepsy and hemiplegia (see below). In any case, it can be observed that strictly mental illnesses were rarely diagnosed among the children admitted to both institutions, mirroring psychiatric theories

[53] Vassiliki Theodorou and Despina Karakatsani, "Eugenic concerns, population policies and puericulture in interwar Greece", *The Historical Review/La revue Historique*, 17 (2020), pp. 53–90. On eugenics in Greece, see Sevasti Trubeta, *Physical Anthropology, Race and Eugenics in Greece 1880s–1970s*, Boston: Brill Academic Publishers, 2013; Alexandra Barmpouti, *Post-War Eugenics, Reproductive Choices and Population Policies in Greece 1950s–1980s*, Cham, Switzerland: Palgrave Macmillan, 2019; Yorgos Kokkinos, *"Άξια" και "Ανάξια" Ζωή. Ευγονική, Εκφυλισμός, Βιοπολιτική: Ο Γιατρός στο Ρόλο του Κοινωνικού Θεραπευτή και του Εθνικού Αναμορφωτή* [*"Worthy" and "Unworthy" Life. Eugenics, Degeneration, Biopolitics: The Doctor in the Role of Social Healer and National Reformer*], Athens: Ταξιδευτής [Taxideftis], 2021.

about the low mental vulnerability of childhood and the unwillingness or inability of the children's environment to recognise their mental problems and seek psychiatric treatment and hospitalisation (see Chapter 6).

Dementia Praecox

The first mention of dementia praecox was made in the mid-nineteenth century by Morel, who called "démence précoce" a condition of severe intellectual impairment and mental and social failing in young people.[54] Its main symptoms were "abnormal" emotions, such as hatred of parents, arrested development, loss of vitality, tendency to isolation, cessation of intellectual faculties and, eventually, dementia.[55] Whereas Morel's "démence précoce" was one of the endpoints of the degenerative process, the "dementia praecox" described by Kraepelin in 1893 was a separate disease, although it also manifested itself as a degenerative process. It appeared abruptly, usually in people up to 25 years of age, and led to a decline in cognitive and especially emotional abilities and energy. Patients were quiet, distracted, listless and expressionless. They isolated themselves, daydreamed and stayed in bed for days or months without feeling the need to do anything. Although they were aware of what was going on around them, they were not affected by it. They were childish in their actions and unable to focus their attention and draw conclusions. Their ideas were well formulated but distorted and disjointed, and they responded slowly and in monosyllables, not because they did not understand what was being said to them, but because they did not want to respond.[56] Another symptom was catatonia, which at the end of the nineteenth century was also considered a separate disease, and was characterised by three disorders of the will: negativism (patients' resistance to what was asked of them and to their own needs and pleasures), stereotypy (maintaining the same position or repeating the same movement, word or phrase) and suggestibility (patient's acceptance of any external influence

[54] Berrios, *The History of Mental Symptoms*, p. 189.

[55] Bénédict-Augustin Morel, "Treatise on mental diseases", in Cousin, Garrabé and Morozou (eds), *Anthology of French Language Psychiatric Texts*, pp. 179–80.

[56] Ion and Beer, "The British reaction to dementia praecox 1893–1913"; Emil Kraepelin, "Lectures on clinical psychiatry" (1894), in Roy Porter (ed.), *The Faber Book of Madness*, London: Faber and Faber, 1991, pp. 393–97.

by remaining in the position they were placed in or by imitating what they saw and heard).[57]

In the period studied, the diagnosis of dementia praecox was used by psychiatrists all over the world, but at the same time it caused intense debate, being considered too broad and vague.[58] In Greece, it was used as early as the 1900s and it was the most common diagnosis in the Eginitio, the Dromokaitio and the Corfu Psychiatric Hospital.[59] In 1906, Miltiadis Ekonomakis, a lecturer in neurology and psychiatry at the University of Athens and resident doctor at the Eginitio from 1904 to 1907, who had studied with Kraepelin in Germany, endorsed the latter's "masterful" description of dementia praecox.[60] Two years later, Katsaras, although following the French school of psychiatry, referred to Kraepelin as the "wise Munich clinician", who had firmly and conclusively established dementia praecox.[61] According to Economakis and Katsaras, the disease

[57] Katsaras, *Pathology of the Nerves and Psychiatry*, v. 2, p. 502; Michael Katsaras, "Περί ψευδών κατατονικών ανοιών. Λόγος εκφωνηθείς τη 20ή Ιανουαρίου 1908 εν τη μεγάλη αιθούση του πανεπιστημίου" ["On pseudo catatonic dementias. Speech delivered on 20 January 1908 in the great hall of the university"], in National University, *Τα κατά την Πρυτανείαν Μιχαήλ Κατσαρά* [*The Activities of the Rectorate of Michael Katsaras*], Athens: Π. Δ. Σακελλαρίου [P.D. Sakellarios Publishing House], 1909, pp. 4–6.

[58] In France, resistance to Kraepelin's classification was also based on non-scientific grounds. The widespread use of dementia praecox represented the expansion of German psychiatry at the expense of French psychiatry, which was not easily accepted, especially at a time of intense anti-German sentiment in France. Dowbiggin, "Back to the future".

[59] Athanasios Karavatos, "Η εισαγωγή στην Ελλάδα του όρου 'πρωτόγονος (ή πρώιμος) άνοια'. Επιδράσεις της γαλλικής και γερμανικής ψυχιατρικής σχολής" ["The introduction in Greece of the term 'primitive (or early) dementia'. Influences of the French and German psychiatric school"], *Ψυχιατρική*, [*Psychiatry*] (1995), pp. 136–44; Dimitris Ploumpidis, *Ιστορία της Ψυχιατρικής στην Ελλάδα. Θεσμοί, Ιδρύματα και Κοινωνικό Πλαίσιο 1850–1920* [*History of Psychiatry in Greece: Institutions, Hospitals and Social Framework 1850–1920*], Athens: Εξάντας [Exandas], 1995, p. 177.

[60] Miltiadis Economakis, "Η πρωτόγονος άνοια (dementia praecox). Κλινική περιγραφή εν συνοδεία ιδίων παρατηρήσεων" ["Dementia praecox. Clinical description accompanied by own observations"] (1906), in Athanasios Karavatos, Dimitris Ploumpidis and Yorgos Christodoulou (eds), *Ανθολόγιο Ελληνικών Ψυχιατρικών Κειμένων* [*Anthology of Greek Psychiatric Texts*], Athens: Ελληνική Ψυχιατρική Εταιρεία [Hellenic Psychiatric Association], 2006, pp. 239–49. On Economakis, see Athanasios Karavatos, Dimitris Ploumpidis and Yorgos Christodoulou, "Οικονομάκης Μιλτιάδης" ["Economakis Miltiadis"], in Karavatos, Ploumpidis and Christodoulou (eds), *Ανθολόγιο Ελληνικών Ψυχιατρικών Κειμένων*, pp. 593–94.

[61] Katsaras, "On pseudo catatonic dementias".

was characterised by a rapid and peculiar impairment of attention, judgement and interest, incoherence and disintegration of thought, excitement, elimination or destruction of the will, stereotypy, suggestibility, and negativism. Patients were indifferent to their moral obligations, derived no pleasure from work, rest or entertainment and usually manifested unstable and incoherent delusions. The diagnosis of pseudo dementia praecox, which was used in the Eginitio, resembled dementia praecox but, unlike it, was curable.[62]

Although it has been argued that dementia praecox was more of a male diagnosis, because its basic features (apathy, senselessness, catatonia) were considered as pathological forms of the male stable nature,[63] the data from the Eginitio and the Dromokaitio do not confirm a link with gender. What they do support is the association of the diagnosis with youth: it was more common in patients aged 14–25. The international expansion of the use of dementia praecox can be interpreted as a facet of the process of establishing adolescence as a distinct age group and as an indicator of the growing psychiatric interest in youth. Indeed, dementia praecox highlights psychiatry's contribution to the discourses that from the last third of the nineteenth century described youth as an important but difficult to control and potentially dangerous segment of society.[64] Clinical descriptions of dementia praecox insisted on issues that were central to such discourses about youth: discipline, morality, sexuality, industriousness, relationships with elders and superiors. Psychiatrists categorised as symptoms of this specific mental illness various signs that the "superficial observer" explained as the result of poor education and behaviour, including disrespect for parents, arrogance, and avoidance of family gatherings.[65] By encompassing these issues in psychiatric categories, psychiatry claimed and, as we shall see in part III of the book, to some extent

[62] Economakis, "Dementia praecox"; Katsaras, "On pseudo catatonic dementias".

[63] Lunbeck, *The Psychiatric Persuasion*, pp. 148–51.

[64] Harry Hendrick, *Images of Youth: Age, Class and the Male Youth Problem, 1880–1920*, Oxford: Clarendon Press, 1990, pp. 3–22, 101–7. Hendrick, using the British case, emphasises that perceptions of the dangers of adolescence in the late nineteenth century were mainly associated with working-class youth, who were more independent than bourgeois youth.

[65] Karl Kahlbaum, "On heboidophrenia" (1890), *History of Psychiatry*, 13 (2002), pp. 201–8 (202).

achieved a share in the definition and management of youth and its disorders.

Dementia, Senile Insanity and Mental Confusion

In the last decades of the nineteenth century, the term dementia had a broad meaning, covering both a psychotic, reversible illness and the irreversible outcome of various mental and nervous diseases.[66] It included a basic set of cognitive symptoms, such as impaired memory, attention and orientation, but also motor disorders, as well as hallucinations, delusions, behavioural problems, apathy, altered personality, mania, and melancholia.[67] Although attempts were made to reduce dementia to an organic and irreparable disorder of intellectual functioning, in the early twentieth century the diagnosis retained its broad meaning and continued to be divided into primary (acute), secondary (chronic) and senile dementia, a distinction made by Esquirol in the early nineteenth century.[68]

Acute or primary dementia was brief and reversible and usually occurred in adults up to the age of 30.[69] In the late nineteenth century, it was included in the diagnosis of mental confusion, which was established in the 1890s and had as its main symptom the absence, confusion or slowing down of ideas.[70] Patients could not express themselves at all, had

66 Henry Maudsley, *The Pathology of Mind: A Study of its Distempers, Deformities, and Disorders*, London: Macmillan, 1995 (first published in 1895), pp. 348–54; German Berrios, "Memory and the cognitive paradigm", in Robin Murray and Trevor Turner (eds), *Lectures on the History of Psychiatry: The Squibb Series*, London: Gaskell, 1990, pp. 194–211; German Berrios, Roy Porter and Tom Kitwood, "Dementia", in German Berrios and Roy Porter (eds), *A History of Clinical Psychiatry: The Origin and History of Psychiatric Disorders*, London: Athlone, 1995, pp. 34–71 (38); Shorter, *A Historical Dictionary of Psychiatry*, p. 268.

67 Berrios, Porter and Kitwood, "Dementia", p. 37; Berrios, *The History of Mental Symptoms*, pp. 178–81; Berrios, "Memory and the cognitive paradigm", pp 194–211; Katsaras, *Pathology of the Nerves and Psychiatry*, v. 2, pp. 270–71, 281–82.

68 Berrios, "Memory and the cognitive paradigm", pp. 194–211; Berrios, Porter and Kitwood, "Dementia", pp. 34, 38; Berrios, *The History of Mental Symptoms*, pp. 183, 200; Shorter, *A Historical Dictionary of Psychiatry*, pp. 71–73; Maudsley, *The Pathology of Mind*, pp. 348–54; Mercier, *A Text-book of Insanity*, p. 293; Régis, *Précis de Psychiatrie*, p. 539.

69 Berrios, *The History of Mental Symptoms*, pp. 175, 180–81.

70 Previously, mental confusion was regarded a symptom of various diseases, such as alcoholism, senile dementia, epilepsy, hysteria and neurasthenia. Philippe Chaslin, *La*

a blank stare, did not care for themselves, lacked emotion, acted incoherently and aimlessly, and made strange and stereotyped movements. They sometimes experienced delusions and hallucinations, as well as physical symptoms such as facial tremors.[71]

In chronic (secondary) and senile dementia, the main symptom was mental impairment, which was the result of attacks of acute insanity (in secondary dementia) or ageing (in senile dementia).[72] Senile dementia usually affected patients aged 65 and over, was incurable and was characterised by lack of accuracy of thought, gaps in memory, especially of recent events, changes in personality and emotions, depression, sometimes psychotic symptoms such as delusions and hallucinations, gradual inability to work and inappropriate behaviour, such as secret love affairs and marriage proposals.[73] The symptoms were largely attributed to organic causes, for example cerebral deterioration and sensory problems, which made it difficult for patients to relate to their environment and adapt to new circumstances, leading them to live in the past. The emergence of senile dementia as a separate disease category should be seen in the context of the process of medicalisation of old age in the second half of the nineteenth century, when, as a result of increased longevity, diseases of the third age became more common and a larger proportion of the elderly were exposed to medical care.[74]

In Greece, based on the evidence from the Eginitio and the Dromokaitio, these processes do not seem to have been underway in the early twentieth century. As noted above (see Chapter 6), patients over the age of 60 were not many in either hospital, which was in line with the rarity of both dementia and non-cognitive diagnoses of old

Confusion Mentale Primitive, Stupidité, Démence Aiguë, Stupeur Primitive, Paris: Asselin et Houzeau, 1895, pp. 123–29; Katsaras, *Pathology of the Nerves and Psychiatry*, v. 2, pp. 468–82; Simonidis Vlavianos, "Η διανοητική πρωτοπαθής σύγχυσις, καθορισμός και συνωνυμία" ["Mental primary confusion, definition and synonymy"], *Ψυχιατρική και Νευρολογική Επιθεώρησις* [*Psychiatric and Neurological Review*], 3 (1904–1905), pp. 43–48.

71 Chaslin, *La Confusion Mentale Primitive*, pp. 10–14, 81, 90–99.

72 Berrios, *The History of Mental Symptoms*, p. 175.

73 Régis, *Précis de Psychiatrie*, pp. 542–44; Maudsley, *The Physiology and Pathology of the Mind*, p. 212; Berrios, *The History of Mental Symptoms*, pp. 180–86.

74 Berrios, *The History of Mental Symptoms*, p. 177; Berrios, "Memory and the cognitive paradigm", pp. 194–211; Berrios, Porter and Kitwood, "Dementia".

age, such as senile melancholia.[75] Dementia was usually conceptualised as senile and organic dementia and was diagnosed in patients aged 50 and over. Other types of dementia (syphilitic, paralytic, alcoholic, toxic, degenerative, and secondary dementia) were even less common than organic and senile dementia and were diagnosed in younger patients with early cognitive impairment.[76] In any case, non-cognitive disorders, such as delusions and mania, continued to be considered key symptoms in patients with dementia, and the diagnosis of mental confusion was often used in conjunction with other, non-cognitive, diagnoses, such as melancholia, alcoholism, and dementia praecox. In other words, the different forms of dementia and the relationship of dementia and mental confusion to non-cognitive symptoms persisted in psychiatric practice in early twentieth-century Greece.

General Paralysis of the Insane (GPI)

General paralysis of the insane was a very common diagnosis in nineteenth- and early twentieth-century asylums in many countries.[77] It was an organic disease with a deteriorating course, almost always ending in death, and characterised by severe mental symptoms.[78] It began with depression, anxiety, memory impairment, fatigue, apathy, confusion, mania, and periodic attacks similar to epileptic seizures.[79] There were also personality changes: indifference to work and family, and interest only in material pleasures such as food. Patients became overly cheerful and made ambitious and irrational plans. They lacked self-control, became selfish and irritable, and behaved foolishly and immorally, ignoring the

[75] Senile melancholia was diagnosed just in 15 patients in the Eginitio, who in Table 8.1 are included in "melancholy". However, the general diagnosis of melancholia was very common in patients 50 years and older at both institutions.

[76] In Tables 8.1 and 8.2, toxic and alcohol dementia have been included in the toxic and alcohol diagnoses, respectively, while paralytic and syphilitic dementia have been included in general paralysis, since, as we will see in the next section, general paralysis was essentially considered a dementia.

[77] Shorter, *A Historical Dictionary of Psychiatry*, p. 193.

[78] Toby Gelfand, "Neurologist or psychiatrist? The public and private domains of Jean-Martin Charcot", *Journal of the History of the Behavioral Sciences*, 36 (2000), pp. 215–29.

[79] Maudsley, *The Pathology of Mind*, pp. 454–57; Yanniris, *The Phrenopathies in Greece*, pp. 48–51.

conventions of decency. Then they became forgetful and manifested delusions of grandeur and hallucinations. In women, the symptoms appeared and developed more rapidly than in men, but were milder, which was attributed to the nobler and gentler female nature and life.[80]

The psychopathological phenomena of GPI were connected with specific brain lesions and this made the disease a nosological and research model in psychiatry, confirming the perception of insanity as a cerebral, organic disease.[81] The organic picture of GPI was reinforced by its physical symptoms (especially disorders of the muscular system),[82] but also by its syphilitic aetiology, well established by the end of the nineteenth century and proven by 1913.[83] At the same time, until the 1920s, psychiatric literature continued to refer to other aetiological factors of GPI that had been identified in the nineteenth century and were related to the temperament of the patients, alcoholic and venereal abuse, city life, immorality and excessive intellectual and moral stress.[84] The proven link between GPI and syphilis confirmed some of these aetiologies, as syphilis was associated with sexual and alcoholic abuse and urban life, and increased anxieties about a perceived epidemic of venereal disease.

Within this context, GPI was strongly associated with life in the city, where the struggle to make a living, intellectual activity, opportunities for abuses and, in the Greek case, "European" habits were thought to be

[80] Katsaras, *Pathology of the Nerves and Psychiatry*, v. 2, pp. 598–600; Simonidis Vlavianos, "Διαφορική διάγνωσις νευρασθενείας και γενικής παραλύσεως μετά πρωτοτύπων κλινικών παρατηρήσεων" ["Differential diagnosis of neurasthenia and general paralysis with original clinical observations"], *Ψυχιατρική και Νευρολογική Επιθεώρησις* [*Psychiatric and Neurological Review*], 2 (1903–1904), pp. 132–35, 165–69, 199–204, 356–60; Maudsley, *The Pathology of Mind*, pp. 436–39, 461, 447–79; Yanniris, *The Phrenopathies in Greece*, pp. 48–51.

[81] Frank Fish, "David Skae, M.D., F.R.C.S, founder of the Edinburgh school of psychiatry", *Medical History*, 9 (1965), pp. 36–53; Klaus Dörner, *Bürger und Irre. Zur Sozialgeschichte und Wissenschaftssoziologie der Psychiatrie*, Frankfurt am Main: Fischer Taschenbuch Verlag, 1975, p. 224; Deniker and Olie, *Fou, Moi?*, pp. 88–9.

[82] Maudsley, *The Pathology of Mind*, pp. 440–49; Vlavianos, "Differential diagnosis of neurasthenia and general paralysis"; Katsaras, *Pathology of the Nerves and Psychiatry*, v. 2, pp. 602–4.

[83] Shorter, *A Historical Dictionary of Psychiatry*, pp. 193–94.

[84] Gayle Davis, *"The Cruel Madness of Love": Sex, Syphilis and Psychiatry in Scotland, 1880–1930*, Amsterdam: Rodopi, 2008, ch. 6.

more intense.[85] In addition, the disease was associated with the lower social strata, which were habitually linked to alcoholism, syphilis and immorality,[86] but it was also diagnosed in the middle and upper classes, to which the other alleged causes of GPI applied, namely the high level of education and strenuous intellectual and professional activity.[87] Based on this epidemiological profile, the disease was diagnosed more often in men, although at the end of the nineteenth century there was an increase in the number of female general paralytics, which psychiatrists attributed to women's greater involvement in the competition of life and the "worries of the world".[88] Still, even when GPI was diagnosed in women, particularly in the upper classes, old age or psychological factors such as bereavement and anxiety were cited as causes.[89] Besides, it is likely that women with GPI avoided hospitalisation for the same reason that women with syphilis did not receive adequate and timely medical treatment: GPI, like syphilis, implied prostitution and carried a greater stigma for women, which led to denial or concealment of the disease.[90]

In the Eginitio and the Dromokaitio, GPI was one of the most common diagnoses, sometimes combined with other diagnoses such as mania, melancholia, and organic encephalopathy. The social characteristics of the general paralytics reflected those pointed out by the psychiatric discourse. They were mostly men but came from all social classes; their occupations ranged from colonels, doctors, and lawyers to clerks, merchants, and workers. Greek psychiatrists argued that in Greece, unlike in other European countries, the rates of female GPI remained low because of national resistance to women's emancipation. The latter, according to Katsaras, was not only incompatible with the nature of women and the ethos of the Greek people, but also dangerous: by freeing

[85] Yeoryios Zilanakis, "Η προϊούσα καθολική παράλυσις. Στατιστική της δεκαπενταετίας 1908–1922" ["General paralysis of the insane. Statistic of the fifteen years 1908–1922"], *Ιατρική Πρόοδος* [*Medical Progress*], 30 (1925), pp. 204–7; Katsaras, *Pathology of the Nerves and Psychiatry*, v. 2, p. 616.

[86] Davis, "*The Cruel Madness of Love*", ch. 6.

[87] Zilanakis, "General paralysis of the insane"; Katsaras, *Pathology of the Nerves and Psychiatry*, v. 2, p. 616.

[88] Maudsley, *The Pathology of Mind*, pp. 466–68; Ellis, "Sex, influence of, in insanity".

[89] Davis, "*The Cruel Madness of Love*", ch. 6.

[90] Jill Harsin, "Syphilis, wives, and physicians: Medical ethics and the family in late nineteenth-century France", *French Historical Studies*, 16 (1989), pp. 72–95.

women from the "golden brake of the family", it weakened them and exposed them to male concerns, abuses and, ultimately, diseases just like GPI.[91]

Nervous Illnesses

Conditions associated with weak, irritable or sensitive nerves begun to occupy a special place in medicine in the seventeenth century, when the English physician Thomas Willis (1621–1675) coined the term "neurology". In the following century, the Scottish nosologist William Cullen (1710–1790) coined the term "neurosis" to describe generalised nervous diseases caused by a dysfunction of the nervous system in the absence of overt organic damage.[92] The distinction between organic and functional nervous diseases was in use in the early twentieth century,[93] with the latter including hysteria, neurasthenia and epilepsy.[94] Both organic and functional nervous diseases were a neurological rather than a psychiatric domain, at least until the end of the nineteenth century.[95] It was at this time that neurologists began to treat functional nervous diseases as mental disorders, because they were still unable to discover anatomical data about these diseases and they observed that hysterical and neurasthenic patients were not cured by electricity, hydrotherapy, special diets or massage, but by the relationship and discussions with the doctor, in

[91] Katsaras, *Pathology of the Nerves and Psychiatry*, v. 2, p. 615.

[92] William Bynum, "The nervous patient in 18th- and 19th-century Britain: The psychiatric origins of British neurology", in Murray and Turner (eds), *Lectures on the History of Psychiatry*, pp. 115–27; Günter Risse, "Hysteria at the Edinburgh Infirmary: The construction and treatment of a disease, 1770–1800", *Medical History*, 32 (1988), pp. 1–22; Shorter, *A Historical Dictionary of Psychiatry*, p. 192; Richard Hunter and Ida Macalpine, *Three Hundred Years of Psychiatry, 1535–1860: A History Presented in Selected English Texts*, Hartsdale, N.Y.: Carlisle Publications, 1982, p. 392.

[93] Edward Shorter, *A History of Psychiatry: From the Era of the Asylum to the Age of Prozac*, New York: John Wiley, 1997, pp. 129–30; Tracey Loughran, "Hysteria and neurasthenia in pre-1914 British medical discourse and in histories of shell-shock", *History of Psychiatry*, 19 (2008), pp. 25–46.

[94] Michael Katsaras, *Παθολογία των Νεύρων και Ψυχιατρική. Νευρώσεις, Νευρίτιδες, Περιφερικαί Παραλύσεις* [*Pathology of the Nerves and Psychiatry. Neuroses, Diseases of the Nerves, Peripheral Paralyses*], v. 3, Athens: Εστία [Estia], 1900, pp. 263 and 307–8.

[95] Ben Shephard, *A War of Nerves*, London: Jonathan Cape, 2000, p. 8.

other words, by "psychotherapy".[96] By the 1900s, part of neurology had moved from neuropathological research to psychotherapy[97] and psychiatrists became increasingly interested in neuroses.[98] During the First World War, doctors were confronted with cases whose symptoms resembled those of hysteria and neurasthenia and were gradually attributed to the emotional stress of war rather than to the physical effects of explosions, gassing or injuries.[99] This experience prompted psychiatry to reconceptualise functional neuroses as mild and borderline forms of mental illness, thus extending its scope to transient and less severe conditions, and consequently to more numerous and wealthy patients than those usually treated in asylums.[100]

In the period under study, however, neurologists continued to play an important role in the treatment of nervous diseases, the meanings of which remained strongly influenced by neurological concepts. It is no coincidence that of the two institutions examined, most nervous diagnoses were recorded at the Eginitio, which was both a neurological and

[96] Janet Oppenheim, *"Shattered Nerves": Doctors, Patients, and Depression in Victorian England*, New York: Oxford University Press, 1991, p. 15; Shorter, *A History of Psychiatry*, pp. 132–39; Simonidis Vlavianos, "Η ψυχοθεραπεία και η επίδρασις αυτής επί των ψυχονευρώσεων" ["Psychotherapy and its effect on psychoneuroses"], *Ψυχιατρική και Νευρολογική Επιθεώρησις [Psychiatric and Neurological Review]*, 2 (1903–1904), p. 321–25; Jules Déjerine, "Μέθοδος της εν νοσοκομείω απομονώσεως δια τας νευρώσεις" ["Method of hospital isolation for neuroses"], *Ψυχιατρική και Νευρολογική Επιθεώρησις* [*Psychiatric and Neurological Review*], 2 (1903–1904), pp. 288–89; Paul Dubois, "Αι ψυχονευρώσεις και η ηθική αυτών θεραπεία" ["Psychoneuroses and their moral treatment"], *Ψυχιατρική και Νευρολογική Επιθεώρησις* [*Psychiatric and Neurological Review*], 2 (1903–1904), pp. 348–56.

[97] Shorter, *A History of Psychiatry*, pp. 136–43. Besides, Freud, the founder of psychoanalysis, was a neurologist.

[98] Jan Goldstein, *Console and Classify: The French Psychiatric Profession in the Nineteenth Century*, Cambridge: Cambridge University Press, 1987, p. 334; Chandak Sengoopta, "'A mob of incoherent symptoms'? Neurasthenia in British medical discourse, 1860–1920", in Marijke Gijswijt-Hofstra and Roy Porter (eds), *Cultures of Neurasthenia from Beard to the First World War*, Amsterdam: Rodopi, 2001, pp. 97–116.

[99] Wendy Holden, *Shell Shock*, London: Channel 4, 1998, pp. 7–19; Jay Winter, "Shell-shock and the cultural history of the Great War", *Journal of Contemporary History*, 35 (2000), pp. 7–11; Paul Lerner, *Hysterical Men: War, Psychiatry, and the Politics of Trauma in Germany, 1890–1930*, Ithaca, N.Y.: Cornell University Press, 2003, p. 78; Shephard, *A War of Nerves*, p. 31.

[100] Shorter, *A Historical Dictionary of Psychiatry*, p. 145; Oppenheim, *"Shattered Nerves"*, p. 9.

Table 8.3 Main types of diagnoses

Diagnoses	*Eginitio*		*Dromokaitio*	
	N	%	*N*	%
Psychiatric	852	59.6	907	94.4
Neurological	414	29.0	18	1.9
Nervous	163	11.4	26	2.7
Unknown	1	0.1	10	1.0
Total	1430	100.0	961	100.0

Source: Sample from the registers, Eginitio and Dromokaitio, 1905–1909 and 1916–1920 (see Table 1.1).

a psychiatric hospital. In addition, a number of neurological conditions were diagnosed almost exclusively at the Eginitio, such as encephalopathy, paralysis, meningitis, and spinal atrophy, which will not be examined here because they had mainly physical symptoms (see Table 8.3). Instead, the analysis is confined to the nervous diseases that had mental symptoms and were found in both institutions.

Hysteria

Hysteria, which had been associated with the uterus since ancient times, began to be understood as a disease of the nervous system in the seventeenth century.[101] By the end of the nineteenth century, the prevailing description of hysteria was largely owed to Jean-Martin Charcot (1825–1893), who saw hysteria as a hereditary disease affecting both men and women and caused by functional abnormalities of the central nervous system.[102] According to Charcot, the main hysterical symptoms, such as paralysis, convulsions, hypersensitivity, and anaesthesia, were neurological in nature. The most spectacular aspect of hysteria, which did not occur in all cases, was the hysterical attack or grand hysteria, characterised by

[101] Katsaras, *Pathology of the Nerves and Psychiatry*, v. 3, pp. 7–8; Shorter, *A Historical Dictionary of Psychiatry*, pp. 133–34; Risse, "Hysteria at the Edinburgh Infirmary"; G.S. Rousseau, "'A strange pathology': Hysteria in the early modern world, 1500–1800", in Sander Gilman, Helen King, Roy Porter, G. S. Rousseau and Elaine Showalter (eds), *Hysteria Beyond Freud*, Berkeley: University of California Press, 1993, pp. 91–221.

[102] Roy Porter, *Madness. A Brief History*, Oxford: Oxford University Press, 2002, p. 139; Shorter, *From Paralysis to Fatigue*, p. 175.

violent movements, intense emotions and epileptoid convulsions.[103] By the end of his life, Charcot seemed to accept a psychological aetiology for hysteria.[104] Indeed, the psychological model of hysteria, which existed since the mid-nineteenth century,[105] began to take hold at the turn of the century, and the psychologisation of hysteria was completed by Freud in the 1920s.[106]

At the beginning of the twentieth century, the basic symptoms of hysteria continued to be the physical disturbances mentioned above. At the same time, mental complications were always described, including hallucinations and agitation, laughing and crying without reason, impatience, a tendency to quarrel or isolate, periodic cheerfulness and unusual talkativeness, or moodiness and depression. Another mental facet of hysteria was the hysterical character, which was perceived as a mild form of the illness and an exaggerated version of the female character: volatile, capricious, enthusiastic, stubborn, quarrelsome, hypocritical, vain, with a tendency to romance, eroticism and lying.[107] In addition, part of the hysterical attack was the hysterical delirium, which was manic, melancholic or hallucinatory. If it lasted long, or if the hysterical attacks were repeated, hysteria was thought to lead to full blown insanity: hysterical insanity took the form of melancholia with stillness and speechlessness or of mania with excitement, laughter, singing, incoherent and rapid speech, often of an

[103] Mark Micale, "Charcot and the idea of hysteria in the male: Gender, mental science, and medical diagnosis in late nineteenth-century France", *Medical History*, 34 (1990), pp. 363–411; Jan Goldstein, "The hysteria diagnosis and the politics of anticlericalism in late nineteenth-century France", *The Journal of Modern History*, 54 (1982), pp. 209–39; Roy Porter, "The body and mind, the doctor and the patient: Negotiating hysteria", in Gilman, King, Porter, Rousseau and Showalter (eds), *Hysteria Beyond Freud*, pp. 225–85; Elaine Showalter, "Hysteria, feminism and gender", in Gilman, King, Porter, Rousseau and Showalter (eds), *Hysteria Beyond Freud*, pp. 286–344; Shorter, *From Paralysis to Fatigue*, pp. 49–50, 69–94.

[104] Shorter, *From Paralysis to Fatigue*, p. 191.

[105] Roy Porter, "Madness and the family before Freud: The view of the mad-doctors", *Journal of Family History,* 23 (1998), pp. 159–72; Joseph Breuer, "Theoretical issues" (1895), in Joseph Breuer and Sigmund Freud, *Studies in Hysteria*, London: Penguin, 2004, pp. 189–254 (190).

[106] Shorter, *A Historical Dictionary of Psychiatry*, pp. 110–11.

[107] Mark Micale, *Approaching Hysteria: Disease and its Interpretations*, Princeton, N.J.: Princeton University Press, 1995, pp. 230–34; Ilias Makris, "Περίπτωσις υστερικών εγκαυμάτων" ["Case of hysterical burns"], *Ιατρική Πρόοδος [Medical Progress]*, 23 (1918), pp. 147–52.

erotic and obscene nature, and inappropriate behaviour, characterised by morbid exhibition of the body, erotic symptoms, feigned physical illness and convulsions.[108]

Most aspects of hysteria, but especially the hysterical character and hysterical insanity, highlight its gendered dimension. Although in the late nineteenth century it could be diagnosed in both sexes, it continued to be understood as a disease mainly of women, especially young women.[109] The female nervous system was considered to be less controlled by the will, less stable and more vulnerable to functional disorders than the male.[110] Moreover, women were thought to be more susceptible to the mental causes of hysteria, such as sudden and intense emotions,[111] not only because they were nervously unstable but also because they were more easily influenced by external impressions and more repressed by social constraints.[112] At the same time, at the turn of the century, hysteria continued to be associated with the female reproductive system and was attributed to irritation of the uterus and ovaries, menstrual irregularities and frequent pregnancies.[113] The tie between hysteria and women was so strong that many descriptions of hysteria were limited to the female gender: hysterical women were so bad-tempered that they made life unliveable for those around them; they were despotically selfish, a scourge on their families and a source of despair for their doctors.[114] Finally, even

[108] Katsaras, *Pathology of the Nerves and Psychiatry*, v. 3, pp. 18–69; 94–96; 174–204; Clouston, *Clinical Lectures*, p. 529; Porter, "The body and mind, the doctor and the patient".

[109] Katsaras, *Pathology of the Nerves and Psychiatry*, v. 3, pp. 10–12.

[110] Carroll Smith-Rosenberg, "The hysterical woman: Sex roles and role conflict in 19th-century America" (1972), in Carroll Smith-Rosenberg, *Disorderly Conduct: Visions of Gender in Victorian America*, New York: A.A. Knopf, 1985, pp. 197–216.

[111] Katsaras, *Pathology of the Nerves and Psychiatry*, v. 3, p. 12.

[112] Horatio Bryan Donkin, "Hysteria" (1892), in Sally Ledger and Roger Luckhurst (eds), *The Fin de Siècle: A Reader in Cultural History, c.1880–1900*, Oxford: Oxford University Press, 2000, pp. 245–50; Thomas Clouston, *The Hygiene of Mind*, London: Methuen & Co, 1906, p. 211.

[113] Smith-Rosenberg, "The hysterical woman", p. 206; Porter, "The body and mind, the doctor and the patient".

Katsaras, *Pathology of the Nerves and Psychiatry*, v. 3, p. 12.

[114] Lynn Gamwell and Nancy Tomes, *Madness in America: Cultural and Medical Perceptions of Mental Illness before 1914*, Binghamton: Cornell University Press, 1995, p. 156.

doctors who argued that hysteria was not just a female disease believed that its symptoms were different in the two sexes. The clinical picture of hysteria in men was fairly stable, with more pronounced physical symptoms, while in women it was fluid, with mainly emotional symptoms such as crying, restlessness, mood swings, and suicide threats.[115]

Neurasthenia

Neurasthenia was described in 1869 by the American neurologist and electrotherapist George Miller Beard (1839–1883) in neurological terms: the central nervous system lost some of its solid components and suffered minor damage to its chemical structure, resulting in a decline in the quality of nerve power.[116] At the beginning of the twentieth century, neurasthenia included a variety of symptoms, such as malaise, fatigue, anorexia and indigestion, visual disturbances, neuralgia, paralysis, insomnia, unwillingness to perform continuous mental work, abnormal menstrual periods in women and sexual disorders in men (reduction of the "sex drive", incomplete erection or total lack of erection, premature ejaculation and spermatorrhoea).[117] In addition to the physical symptoms, an important part of neurasthenia was undue anxiety and despair and phobias,[118] the most common of which was nosophobia: neurasthenics were hypochondriacs, namely they exaggerated about even the slightest health problem, seeking help from several doctors. Finally, the character of the neurasthenics was constantly changing and showed abnormalities such as bad temper, irritability, dullness, melancholy, discouragement, apathy, indifference and a tendency to isolation.[119]

[115] Micale, "Charcot and the idea of hysteria in the male".

[116] Shorter, *A Historical Dictionary of Psychiatry*, p. 187.

[117] Ibid., pp. 187–88; Shorter, *A History of Psychiatry*, p. 130; Gilles de La Tourette, *Les États Neurasthéniques: Formes Cliniques, Diagnostic, Traitement*, Paris: J.-B. Baillière et fils, 1898, pp. 10–25; Vlavianos, "Differential diagnosis of neurasthenia and general paralysis"; Katsaras, *Pathology of the Nerves and Psychiatry*, v. 3, pp. 291–300; S. Zografidis, "Η ψυχική κατάστασις των νευρασθενικών" ["The mental state of the neurasthenics"], *Ιατρική Πρόοδος* [*Medical Progress*], 5 (1900), p. 303.

[118] De La Tourette, *Les États Neurasthéniques*, pp. 28–32.

[119] Vlavianos, "Differential diagnosis of neurasthenia and general paralysis"; M. André-Thomas, "Θεραπεία της υστερίας και της νευρασθενείας δια της απομονώσεως και της ψυχοθεραπείας" ["Treatment of hysteria and neurasthenia through isolation and psychotherapy"], *Ψυχιατρική και Νευρολογική Επιθεώρησις* [*Psychiatric and Neurological*

The period from the late nineteenth century to the 1920s is considered the golden age of neurasthenia,[120] with neurologists and psychiatrists arguing that neurasthenia had become endemic in some countries, including Greece.[121] The popularity of the diagnosis can be attributed to the growing anxiety about modern life at the fin de siècle, as well as the fact that it combined in one organic, nerve-related label a variety of elusive conditions that caused widespread concern and supplied lucrative private clinics with clients.[122] Patients, for their part, also appropriated the diagnosis of neurasthenia, because it explained their problems in biological terms, but also because it had positive socio-economic connotations, being associated mainly with the wealthy and educated bourgeoisie.[123] Nevertheless, at the beginning of the twentieth century, many doctors contended that people of all classes could suffer from neurasthenia.[124]

The expansion of the diagnosis concerned not only social class but also gender. Whereas according to Beard neurasthenia was a male condition, in the early twentieth century it was attributed to women too, but not equally, as women were thought to be protected from the mental and

Review], 2 (1903–1904), pp. 336–48; Katsaras, *Pathology of the Nerves and Psychiatry*, v. 3, p. 293; Zografidis, "The mental state of the neurasthenics".

120 Marijke Gijswijt-Hofstra, "Introduction", in Gijswijt-Hofstra and Porter (eds), *Cultures of Neurasthenia*, pp. 1–30; Tom Lutz, "Varieties of medical experience: Doctors and patients, psyche and soma in America", in Gijswijt-Hofstra and Porter (eds), *Cultures of Neurasthenia*, pp 51–76; Mathew Thomson, "Neurasthenia in Britain: An overview", in Gijswijt-Hofstra and Porter (eds), *Cultures of Neurasthenia*, pp. 77–96; Brad Campbell, "The making of 'American': Race and nation in neurasthenic discourse", *History of Psychiatry*, 2007 (18), pp. 157–78.

121 André-Thomas, "Treatment of hysteria and neurasthenia through isolation and psychotherapy"; Simonidis Vlavianos, "Εθνική νευρασθένεια" ["National neurasthenia"], *Ψυχιατρική και Νευρολογική Επιθεώρησις* [*Psychiatric and Neurological Review*], 6 (1907–1908), pp. 3–5.

122 Gijswijt-Hofstra, "Introduction", p. 26; Doris Kaufmann, "Neurasthenia in Wilhelmine Germany: Culture, sexuality and the demands of nature", in Gijswijt-Hofstra and Porter (eds), *Cultures of Neurasthenia*, pp. 161–76.

123 Thomson, "Neurasthenia in Britain", pp. 77–96; Showalter, "Hysteria, feminism and gender", pp. 294–95; Shorter, *A History of Psychiatry*, p. 130; Gijswijt-Hofstra, "Introduction", p. 1.

124 Gijswijt-Hofstra, "Introduction", p. 23.

moral strain associated with neurasthenia.[125] It was argued, however, that the changing female condition, which caused greater physical stress and nervous irritation, made women vulnerable to nervous exhaustion.[126] As a result, neurasthenia was presented, along with hysteria and migraine, as the scourge of modern city women.[127]

Epilepsy

The case of epilepsy was quite different from that of hysteria and neurasthenia, as research into its pathological anatomy was more successful. For partial epilepsy (Bravais-Jackson epilepsy), brain localisations were discovered in the second half of the nineteenth century, and at the beginning of the twentieth century the hope of similar discoveries for common epilepsy had not been abandoned, as had been the case for hysteria and neurasthenia.[128] The neurological studies of epilepsy increasingly questioned its relationship with insanity, but psychiatry was still concerned with epilepsy:[129] it was considered to have a common pathology and aetiology with mental illness; to be the stage of degeneration prior to insanity;[130] and, above all, to be characterised by specific mental disorders in addition to the classic epileptic symptoms, i.e. convulsive seizures and loss or disturbances of consciousness.[131]

The mental symptoms of epilepsy were organised under the categories of epileptic character and epileptic insanity.[132] The former was

[125] Ibid., p. 23–24; de La Tourette, *Les États Neurasthéniques*, p. 9; Anon., "Διάγνωσις της νευρασθενείας" ["Diagnosis of neurasthenia"], *Ιατρική Πρόοδος* [*Medical Progress*], 20 (1915), pp. 95–96; Katsaras, *Pathology of the Nerves and Psychiatry*, v. 3, p. 299.

[126] Clouston, *The Hygiene of the Mind*, pp. 210–11.

[127] Thomas Clouston, "Developmental insanities and psychoses. The delirium and night terrors, &c., of children. The insanities of children. The insanities of puberty and adolescence", in Daniel Hack Tuke (ed.), *A Dictionary of Psychological Medicine: Giving the Definition, Etymology and Synonyms of the Terms used in Medical Psychology with the Symptoms, Treatment, and Pathology of Insanity and the Law of Lunacy in Great Britain and Ireland*, v. 1, London: J. & A. Churchill, 1892, pp. 357–71.

[128] Katsaras, *Pathology of the Nerves and Psychiatry*, v. 2, pp. 102–13 and v. 3, pp. 235, 253–54.

[129] Hunter and Macalpine, *Three Hundred Years of Psychiatry*, pp. 1044–48.

[130] Morel, "Treatise on mental diseases", pp. 176–80.

[131] Maudsley, *The Pathology of Mind*, p. 480; Katsaras, *Pathology of the Nerves and Psychiatry*, v. 3, pp. 235, 241–43.

[132] Katsaras, *Pathology of the Nerves and Psychiatry*, v. 2, ch. 5.

the distinctive mental state of epileptics between epileptic seizures and was described as changeable, irritable, moody, unpredictable, resentful, hypocritical, opinionated, excessively religious or morally perverse with a tendency to crime.[133] The same symptoms were repeated in epileptic insanity, which was periodic and melancholic, delusional or, more often, manic. It was characterised by delusions, hallucinations, violence, and aggression. As a result, individuals with epileptic insanity were considered to be particularly dangerous and prone to crime. They acted automatically and unconsciously and subsequently they forgot what they had done. Chronic epileptic insanity usually led to epileptic dementia, which was also marked by hallucinations, agitation, violence and dangerousness. Psychiatric analyses of epileptic insanity had a strong legal content because of the alleged propensity of such patients to commit crimes and the forensic problems posed by "mental automatism" and "suspended consciousness".[134]

Other Nervous Illnesses

Apart from hysteria, neurasthenia and epilepsy, other nervous illnesses were occasionally diagnosed in the two institutions. In the Eginitio, a few patients were diagnosed with hypochondria, a condition of lethargy and marked concern for one's health.[135] Hypochondria was a popular diagnosis in the eighteenth century, mainly for men of the upper social strata, but in the second half of the nineteenth century it was largely replaced by neurasthenia.[136] In the early twentieth century, hypochondria was mainly

[133] Maudsley, *The Pathology of Mind*, pp. 486–88; Katsaras, *Pathology of the Nerves and Psychiatry*, v. 3, p. 251; Michael Yanniris, *Περί της Νοσηλείας και Περιθάλψεως των Φρενοπαθών εν τοις Φρενιατρείοις ή κατ' Οίκον: Προς Μόρφωσιν Νοσοκόμων* [*On the Care and Treatment of Insane Persons in Lunatic Asylums or at Home: For the Education of Nurses*], Athens: s.n, 1903, pp. 13–14; German Berrios and Roy Porter, "Epilepsy", in Berrios and Porter (eds), *A History of Clinical Psychiatry*, pp. 147–73.

[134] Maudsley, *The Pathology of Mind*, pp. 477–88; Craig, *Psychological Medicine*, pp. 267–69.

[135] Shorter, *A Historical Dictionary of Psychiatry*, p. 132.

[136] Showalter, "Hysteria, feminism and gender", pp. 294–95.

a symptom of other diagnoses, such as melancholia and neurasthenia,[137] although it continued to be considered as a separate "neuropathy".[138]

Finally, it is worth noting that one army patient in the Eginitio sample was diagnosed with "traumatic hysteria", one of the diagnoses—the most famous being "shell shock"—created during the First World War to group together symptoms observed in the armed forces, such as temporary loss of sight, hearing and voice, stuttering, paralysis, tremors, tics, exhaustion, nightmares and insomnia, memory loss, depression, despair, anxiety, fears, apathy, hallucinations, irritability, inability to concentrate, mental confusion, and epileptoid seizures.[139] "War neuroses" were rare in the Eginitio and the Dromokaitio, although patients of the armed forces or "service patients", as they were called, were treated there. Some of these patients were diagnosed with nervous diseases, but not specifically "war neuroses", while most were found to be suffering from typical mental illnesses of the time, such as general paralysis, dementia praecox, mental confusion, melancholia, and mania. The fact that "service patients" were primarily diagnosed with mental illnesses, which were usually deemed more serious than functional nervous illnesses, suggests that military hospitals transferred cases to psychiatric institutions that were considered more serious and less likely to recover quickly.[140] Indeed, the case histories of the "service patients" emphasise the intense mental symptoms that these patients had manifested in the military hospitals (see Chapter 10).

Overall, nervous illnesses were diagnosed in a minority of patients in the Eginitio and the Dromokaitio (Tables 8.1, 8.2, and 8.3), but there

[137] Gilbert Ballet, *Psychoses et Affections Nerveuses*, Paris: O. Doin, 1897, pp. 138–46.

[138] G. Price, "Αιτιολογία των μη οργανικών φρενοπαθειών" ["Aetiology of the non-organic phrenopathies"], *Ιατρική Πρόοδος* [*Medical Progress*], 17 (1912), pp. 55–58.

[139] Mark Micale and Paul Lerner, "Trauma, psychiatry, and history: A conceptual and historiographical introduction", in Mark Micale and Paul Lerner (eds), *Traumatic Pasts: History, Psychiatry, and Trauma in the Modern Age, 1870–1930*, Cambridge: Cambridge University Press, 2001, pp. 1–30; Andreas Killen, *Berlin Electropolis: Shock, Nerves, and German Modernity*, Berkeley: University of California Press, 2006, p. 130; Holden, *Shell Shock*, pp. 7–11; Craig, *Psychological Medicine*, pp. 305–8; Konstantinos Papastratiyakis, "Τα νευρικά και ψυχικά υπολείμματα των τραυματισμών του κρανίου" ["The nervous and mental remnants of skull injuries"], *Ιατρική Πρόοδος* [*Medical Progress*], 14 (1927), pp. 369–80.

[140] This was also a common practice in other countries, such as Britain. R. Eager, "A record of admissions to the mental section of the Lord Derby War Hospital, Warrington, from June 1916 to June 16th 1917", *Journal of Mental Science*, 64 (1918), pp. 272–96.

was a noticeable difference in the meanings of these diagnoses in each institution. In the Eginitio, which was a psychiatric and a neurological hospital and where neurological diagnoses were more common than in the Dromokaitio, nervous illnesses were understood mostly in neurological terms. A telling example is epilepsy, as it was diagnosed with a similar frequency in both hospitals (2.5%). However, in the Dromokaitio it was more likely to present mental symptoms, whereas in the Eginitio it was usually Bravais-Jackson epilepsy with neurological symptoms such as convulsions, loss of consciousness and muscle rigidity. In addition, in the Dromokaitio, hysteria and neurasthenia were not only very rare but were mainly understood in psychiatric terms: they were diagnosed as "hysterical psychopathy" and "neurasthenic psychosis", respectively. In other words, the meanings of nervous illnesses in Greek neuropsychiatric practice highlight the ambiguities and blurred boundaries between neurology and psychiatry, neurosis and psychosis.

Regarding the social characteristics of the "nervous" patients in the two hospitals, epilepsy was not tied to a particular social class or gender. Quite the opposite, the social physiognomy of the hysterics and neurasthenics was more specific. These diagnoses followed the distinction between female hysteria and male neurasthenia, although not entirely. Besides, the director of the Eginitio endorsed Charcot's view that men could also suffer from hysteria.[141] Finally, the occupations of the patients seem to confirm the tendency of neuropsychiatrists to use neurasthenia for higher social classes than hysteria. In the Eginitio, where the two diagnoses are more frequent and therefore the data more abundant, hysteria was more often diagnosed than neurasthenia in patients with lower middle-class and working-class occupations, such as petty traders, low-ranking clerks, maids, and workers.

Alcoholic and Toxic Diagnoses

Nineteenth- and twentieth-century psychiatric theory closely linked toxic substances, mainly alcohol but also opium, cannabis, lead and others, with mental illness. Alcohol was seen not only as causing or exacerbating mental illness,[142] but also as affecting the drinkers' offspring,

[141] Katsaras, *Pathology of the Nerves and Psychiatry*, v. 3, p. 9.

[142] Clouston, *Clinical Lectures*, p. 481; Maudsley, *The Pathology of Mind*, p. 492; Valentin Magnan, *On Alcoholism: The Various Forms of Alcoholic Delirium and their*

who were thought to be more likely to drink and to become immoral, criminal and insane.[143] Some psychiatrists in the late nineteenth century argued that a quarter of patients in asylums were insane because of alcohol abuse or alcoholic heredity:[144] moreover, by the end of the eighteenth century the concept of drunkenness itself as a mental illness had developed,[145] and in the nineteenth century specific alcoholic diagnoses were described, such as monomania of inebriation ("monomanie d'ivresse")[146] and dipsomania,[147] in which the desire for alcohol was irresistible and self-control was abolished. In the mid-nineteenth century, the term "alcoholism" was coined to group a variety of physical, mental

Treatment, transl. W. S. Greenfield, London: H. K. Lewis, 1876, pp. 242–9; Yanniris, *The Phrenopathies in Greece*, p. 40; Katsaras, *Pathology of the Nerves and Psychiatry*, v. 2, p. 643; Yanniris, "On diseases caused by alcohol in Greece"; Simonidis Vlavianos, "Περί των εκ των οινοπνευματωδών παθήσεων εν Ελλάδι" ["On the diseases caused by alcohol in Greece"], *Ψυχιατρική και Νευρολογική Επιθεώρησις [Psychiatric and Neurological Review]*, 1 (1902–1903), pp. 333–36.

143 Margaret Thomson, "The wages of sin: The problem of alcoholism and general paralysis in nineteenth-century Edinburgh", in William Frederick Bynum, Roy Porter and Michael Shepherd (eds), *The Anatomy of Madness: Essays in the History of Psychiatry*, v. 3, London: Tavistock, 1988, pp. 316–40; Stephen Snelders, Frans J. Meijman and Toine Pieters, "Heredity and alcoholism in the medical sphere: The Netherlands, 1850–1900", *Medical History*, 51 (2007), pp. 219–36.

144 Claude Quétel and Jean-Yves Simon, "L' aliénation alcoolique en France (XIXe siècle et 1ère moitié du XXe siècle)", *Histoire, Economie & Société*, 7 (1988), pp. 507–33; Achilleas Armodios, "Ο αλκοολισμός υπό υγιεινήν και κοινωνιολογικήν έποψιν", ["Alcoholism from a hygienic and sociological point of view"], *Ψυχιατρική και Νευρολογική Επιθεώρησις* [*Psychiatric and Neurological Review*], 2 (1903–1904), pp. 14–23, 40–48, 71–80; Yanniris, "On diseases caused by alcohol in Greece"; Simonidis Vlavianos, "Ο αλκοολισμός", *Ψυχιατρική και Νευρολογική Επιθεώρησις*, 8: 5–6 (1909–1910), pp. 3–192.

145 Matthew Warner Osborn, "Diseased imaginations: Constructing delirium tremens in Philadelphia, 1813–1832", *Social History of Medicine*, 19 (2006) pp. 191–208; J. S. Madden, "Substance use disorders: Clinical section", in Berrios and Porter (eds), *A History of Clinical Psychiatry*, pp. 656–67; Hunter and Macalpine, *Three Hundred Years of Psychiatry*, p. 587; Roy Porter, "The drinking man's disease: The 'pre-history' of alcoholism in Georgian Britain", *British Journal of Addiction*, 80 (1985), pp. 385–96.

146 Étienne Esquirol, *Des Maladies Mentales: Considérées sous les Rapports Médical, Hygiénique et Médico-Légal*, Paris: J.-B. Baillière, 1858, v. 2, pp. 72–83.

147 Maudsley, *The Pathology of Mind*, pp. 500–1; Magnan, *On Alcoholism*, p. 238; Mariana Valverde, *Diseases of the Will: Alcohol and the Dilemmas of Freedom*, Cambridge: Cambridge University Press, 1998, p. 48. As noted in the section on degeneration, Greek psychiatrists included dipsomania in the impulses that characterised the degenerates.

and moral phenomena attributed to chronic alcoholic intoxication,[148] but also to denote a chronic insanity, characterised by progressive physical and mental impairment and mental and moral dullness.[149] Chronic alcoholics suffered from motor and speech disorders, poor concentration and memory, agitation or depression, apathy and irritability, frightening hallucinations, and delusions that could lead to violent and criminal acts. They were also unreliable, suspicious, and uncontrollable; they misunderstood and violated moral and social rules; they could not work systematically; they were indifferent to their family, caused family distress, upheavals, and separation, and blamed others for their own suffering. Eventually, the illness resulted in dementia.[150]

In the second half of the nineteenth century, acute conditions were included in the concept of alcoholism: acute alcoholism or alcoholic delirium or simply drunkenness described the transition from the elation of drinking to agitation, mental confusion, paralysis, unconsciousness and coma. In the most intense forms, there was a change of character with restless depression, suspicion, nervous excitement and nightmares, diminished willpower, intellectual deterioration, and moral apathy. After repeated abuses, the major alcoholic delirium occurred, an insanity with delusions, mania, melancholia or cataplexy and intense, paradoxical,

[148] Henri Bernard, "Alcoolisme et antialcoolisme en France au XIXe siècle: Autour de Magnus Huss", *Histoire, Economie & Société*, 3 (1984), pp. 609–28; Quétel and Simon, "L' aliénation alcoolique en France"; Brian Harrison, *Drink and the Victorians: The Temperance Question in England, 1815–1872*, London: Faber and Faber, 1971, p. 21; Simon Apostolidis, "Ο αλκοολισμός εν Κωνσταντινοπόλει παρά τοις ομογενέσι κατοίκοις", ["Alcoholism in Constantinople among the Greek community"], *Ψυχιατρική και Νευρολογική Επιθεώρησις* [*Psychiatric and Neurological Review*], 6 (1907–1908), pp. 275–84, 335–43.

[149] On the introduction and conceptualisation of "alcoholism" in Greece, and more generally on the ways in which Greek doctors and psychiatrists perceived alcohol consumption in the last quarter of the nineteenth century and the first decades of the twentieth century, see Kostis Gotsinas, *Ένα Τελευταίο Ποτήρι; Αλκοόλ και Αντιαλκοολισμός στην Ελλάδα (1870–1940)* [*One Last Cup? Alcohol and Anti-Alcoholism in Greece (1870–1940)*], Research Centre for the Humanities Digital Library, 2020, pp. 9–11, 5, 12–13, 16; Kostis Gotsinas, "'Disciples of Asclepius' or 'advocates of Hermes'? Psychiatrists and alcohol in early-twentieth-century Greece", in Waltraud Ernst and Thomas Müller (eds), *Alcohol, Psychiatry and Society*, Manchester: Manchester University Press, 2022, pp. 220–47.

[150] Magnan, *On Alcoholism*, pp. 154–56, 160–176; Katsaras, *Pathology of the Nerves and Psychiatry*, v. 2, pp. 650–54; Craig, *Psychological Medicine*, pp. 202–7; Maudsley, *The Pathology of Mind*, pp. 495–500.

changeable and usually frightening hallucinations.[151] Acute alcoholism was subdivided into two types: delirium tremens and alcoholic mania (mania a potu). The former was accompanied by fever and uncontrollable, generalised body tremor and incomplete paralysis of the limbs.[152] In alcoholic mania, the patient had delusions of persecution or sometimes grandeur, became suspicious, uncontrollable, agitated and violent.[153] In general, these patients were dangerous, cunning and obscene and committed inappropriate or criminal acts, which often resulted to their commitment to an institution.[154]

Discussions about alcohol, especially from the second half of the nineteenth century, were part of a wider medical discourse on intoxication or poisoning from toxic substances such as lead, opium, morphine, cocaine, caffeine, and nicotine.[155] These substances were problematised in medical thinking by analogy with alcohol, to which they were claimed to have similar effects.[156] On the one hand, mental states resulting from the abuse (or unintentional use) of toxic substances were grouped under acute and chronic toxic insanities, which were characterised by similar symptoms to alcoholic insanities, mainly by multiple hallucinations that

151 Magnan, *On Alcoholism*, pp. 5, 32–34, 44; Κατσαράς, *Pathology of the Nerves and Psychiatry*, v. 2, pp. 640–41; Maudsley, *The Pathology of Mind*, pp. 492–94; Craig, *Psychological Medicine*, p. 198.

152 Osborn, "Diseased imaginations"; Magnan, *On Alcoholism*, pp. 37–8; Katsaras, *Pathology of the Nerves and Psychiatry*, v. 2, pp. 646–47.

153 Maudsley, *The Pathology of Mind*, pp. 494–96; Clouston, *Clinical Lectures*, pp. 489–90; Craig, *Psychological Medicine*, pp. 201–2.

154 Michael Yanniris, "Περί των εξ οινοπνεύματος παθήσεων εν Ελλάδι" ["On diseases caused by alcohol in Greece"], *Ιατρική Πρόοδος* [*Medical Progress*], 8 (1903), pp. 179–83.

155 Bernard, "Alcoolisme et antialcoolisme", pp. 609–28; Valverde, *Diseases of the Will*, p. 50; William Bynum, "Alcoholism and degeneration in 19th-century European medicine and psychiatry", *British Journal of Addiction*, 79 (1984), pp. 59–70; Terry Parssinen and Karen Kerner, "Development of the disease model of drug addiction in Britain, 1870–1926", *Medical History*, 24 (1980), pp. 275–96.

156 Armodios, "Alcoholism from a hygienic and sociological point of view"; Morel, *Traité des Dégénéréscences*, p. 48; Valverde, *Diseases of the Will*, pp. 50–59; Patricia Prestwich, "Traiter les alcooliques à la Belle Époque: Le rêve du 'grand renfermement'", in Francois-Olivier Touati (ed.), *Maladies, Médecines et Sociétés: Approches Historiques pour le Present*. Actes du VIe Colloque, v. 2, Paris: L' Harmattan, 1993, pp. 218–23; Jean-Jacques Yvorel, "Les mots pour le dire: Naissance du concept de toxicomanie", in Francois-Olivier Touati (ed.), *Maladies, Médecines et Sociétés: Approches Historiques pour le Present*. Actes du VIe Colloque, v. 1, Paris: L'Harmattan, 1993, pp. 209–17.

provoked fear and led to melancholic or manic states.[157] On the other hand, from the late 1870s, substance abuse itself was transformed into a mental illness along the lines of the monomania of inebriation and dipsomania. This gave rise to the concept of morphine addiction, followed by cocaine addiction, opium addiction and the overarching concept of drug addiction.[158]

More specifically, morphine addiction was characterised by a weakening of the character, energy and will: morphine addicts could not act in their own interests and perform their duties, were agitated, lost their memory and became sluggish, irritable, moody and restless. They also lied, were secretive and even went as far as committing crimes to obtain morphine because of the intensity of withdrawal symptoms, such as neuralgia, sweating, hallucinations, delusions, insomnia, confusion, and weakness. Eventually, morphine addicts were driven to mental and physical collapse and even death by wasting or heart paralysis. As for cocaine, in small doses it brought mild stimulation and a feeling of increased strength, while in large doses it reduced mental and physical strength and produced acute delusional states. Patients became talkative, irritable, dangerous to themselves and others, and suspicious; they had delusions of persecution, epileptoid attacks and more intense hallucinations than those of alcoholics.[159] Cannabis induced stupor, hallucinations, agitation, inappropriate behaviour, impaired concentration and actual insanity, manifested by delusions, homicidal and suicidal tendencies and loss of memory and energy.[160] In Greece, doctors claimed that cannabis use was more common than that of morphine and cocaine and spread across

[157] Katsaras, *Pathology of the Nerves and Psychiatry*, v. 2, p. 632.

[158] Parssinen and Kerner, "Development of the disease model of drug addiction"; Nikolaos Arkalidis, "Οι τοξικομανείς εν Ελλάδι" ["Drug addicts in Greece"], in *Πανηγυρικόν Τεύχος επί τη Τριακονταπενταετηρίδι Μιχαήλ Κατσαρά: Εκδοθέν υπό Συναδέλφων και Μαθητών αυτού, 29 Μαρτίου 1928* [*Commemorative Issue on the Thirty-fifth Anniversary of Michael Katsaras: Published by his Colleagues and Students, 29 March 1928*], Athens: Π. Δ. Σακελλάριος [P.D. Sakellarios Publishing House], pp. 91–97.

[159] Katsaras, *Pathology of the Nerves and Psychiatry*, v. 2, pp. 668–71; Craig, *Psychological Medicine*, pp. 214–19.

[160] Thomas Clifford Allbutt and Walter Ernest Dixon, "Opium poisoning and other intoxicants", in Thomas Clifford Allbutt and Humphry Davy Rolleston (eds), *A System of Medicine by Many Writers*, London, 1906, pp. 965–68; John Warnock, "Insanity from hasheesh", *Journal of Mental Science*, 49 (1903), pp. 96–110.

all social classes, although it was mainly associated with the lower social strata, for example soldiers, prisoners and "vagrants".[161]

It becomes obvious that psychiatric descriptions of alcoholic and toxic insanities insisted on their economic, social and moral aspects, adopting and reinforcing their negative popular representations. Although psychiatrists advocated the treatment of alcoholics as sick people,[162] they continued to perceive alcoholism as an immoral and anti-social behaviour.[163] Alcoholic and toxic insanities caused by use rather than unintentional intoxication were inextricably linked with crime, idleness, poverty, "moral ruin", sexual promiscuity, the destruction of family happiness, divorce, and suicide, but also with national mental and physical decline,[164] since alcohol and toxic substances were regarded as "racial poisons".[165] Psychiatrists established a fertile field for public discourse and social intervention based on toxic conditions, particularly in the case of alcohol, which was considered more pervasive than the other intoxicants. In fact, psychiatrists not only fed the anti-alcohol discourse of their time with arguments, but also participated in anti-alcohol organisations and in efforts to establish asylums for alcoholics and to enact anti-alcohol legislation with the stated aim of protecting public health.[166]

161 Arkalidis, "Drug addicts in Greece".

162 Simonidis Vlavianos, "Ο αλκοολισμός και αι ολέθριαι αυτού συνέπειαι" ["Alcoholism and its disastrous consequences"], *Ψυχιατρική και Νευρολογική Επιθεώρησις* [*Psychiatric and Neurological Review*], 6 (1907–1908), pp. 97–128.

163 Thomson, "The wages of sin", pp. 316–40; Nikos Tzavaras, "Κοινωνιολογικές και ψυχιατρικές διαστάσεις του αλκοολισμού" ["Sociological and psychiatric dimensions of alcoholism"], *Σύναψις* [*Synapsis*], 1 (2005), pp. 64–72.

164 Katsaras, *Pathology of the Nerves and Psychiatry*, v. 2, pp. 635–39; Simonidis Vlavianos, "Ο αλκοολισμός" ["Alcoholism"], *Ψυχιατρική και Νευρολογική Επιθεώρησις* [*Psychiatric and Neurological Review*], 8 (1909–1910), pp. 3–192 (157–59); Armodios, "Alcoholism from a hygienic and sociological point of view"; Craig, *Psychological Medicine*, p. 214.

165 On the perceived national and racial impact of alcohol and drugs as factors of degeneration, see Bynum, "Alcoholism and degeneration"; Kostis Gotsinas, *Επί της Ουσίας. Ιστορία των Ναρκωτικών στην Ελλάδα (1875-1950)* [*In Substance. History of Drugs in Greece (1875–1950)*], Heraclio: Πανεπιστημιακές Εκδόσεις Κρήτης [Crete University Press], 2021, pp. 302–5.

166 Bynum, "Alcoholism and degeneration"; Osborn, "Diseased imaginations"; Simonidis Vlavianos, "Ο ανταλκοολικός αγών εν Ελλάδι" ["The anti-alcoholic campaign in Greece"], *Ψυχιατρική και Νευρολογική Επιθεώρησις* [*Psychiatric and Neurological Review*], 8 (1909–1910), pp. 1–2; Gotsinas, *One Last Cup?*, pp. 11–6. Vlavianos, for example,

The psychiatric discourse on alcohol and substance use distinguished between men and women. The latter were thought as lighter users and as being less threatened, because their brain was less affected by such organic agents.[167] On the contrary, some psychiatrists argued that women succumbed to the effects of substances more quickly and were more damaged than men.[168]In addition, psychiatrists differentiated the symptoms of toxic insanities in the two sexes: in alcoholic insanities, for example, women were more likely than men to manifest delusions of jealousy because they were more emotional and devoted exclusively to the household.[169] Finally, at the turn of the century, psychiatrists argued that the habit of using alcohol and other substances was becoming more common among middle-class women, and that the number of women with alcoholic insanities was swelling because women were increasingly exposed to the pressures and harmful habits of the public life.[170] In Greece, psychiatrists were confident that the Greek woman was a rare "victim of alcoholism", but warned that emancipation and changes in lifestyle would lead to an increase in alcoholism among women.[171]

In the Eginitio and the Dromokaitio, alcoholic and toxic diagnoses, such as alcoholism, alcoholic or toxic insanity, delirium tremens, dipsomania, morphine and cocaine mania, were given to a small proportion of patients, mainly men. Yet, toxic substances, and especially alcohol, were reported in many more patients' records as the cause of various illnesses. Therefore, a key driver of admission to the institutions was not substance use, abuse or addiction but the manifestation of severe and chronic symptoms caused by alcohol and toxic substances. As the Greek psychiatrists of the time noted, acute alcoholics, that is, drunks, were rarely admitted to asylums, while chronic alcoholics were only hospitalised when they exhibited loud and persistent delusions[172] or when they were in an advanced

called for the establishment of an anti-alcohol association (Vlavianos, "On the diseases caused by alcohol in Greece").

167 Ellis, "Sex, influence of, in insanity".

168 Clouston, *The Hygiene of the Mind*, pp. 209–10.

169 Katsaras, *Pathology of the Nerves and Psychiatry*, v. 2, p. 653.

170 Ellis, "Sex, influence of, in insanity".

171 Yanniris, "On diseases caused by alcohol in Greece"; Vlavianos, "Alcholism".

172 Vlavianos, "Alcoholism".

stage of mental and physical decline.[173] If the transformation of alcoholism and drug addiction into mental illnesses brought new patients under the psychiatric jurisdiction,[174] the Eginitio and the Dromokaitio do not appear to have participated in this expansion of the psychiatric field.

Symptoms: The "Present Condition" of the Patients

Although the process of psychiatric diagnosis took into account the subtle nuances, the manner and sequence of manifestation, the combination and the causes of symptoms, the above analysis has highlighted the common core symptoms that the diagnoses used at the Eginitio and the Dromokaitio identified and grouped together: agitation, depression, delusions, hallucinations, mental and physical weakness, and bizarre, nonsensical and inappropriate speech and behaviour. Similar symptoms were recorded in the patients' case histories, specifically in the "present condition" section, where the doctors reported their own observations of the patients on admission to the hospital. In the remainder of this chapter, I summarise these symptoms, an approach which has the disadvantage of obscuring the patients' personal history and the combination and variation of their problems, but presents the advantage of offering a concise overview of the most common problems identified by the doctors.

For psychiatrists, the "symptom" constituted a clinical entity that could be objectively observed, described and evaluated, although not always easily or satisfactorily, as patients were often unwilling or unable to cooperate during psychiatric examination, while third party information was not always sufficient or reliable.[175] In this sense, the identification of symptoms was a process of interpreting the patients' speech, behaviour, and appearance (primarily facial expressions, gaze and demeanour), during which the doctors placed particular emphasis on the patients'

[173] Yanniris, "On diseases caused by alcohol in Greece".

[174] Prestwich, "Drinkers, drunkards, and degenerates".

[175] Clouston, *Clinical Lectures*, p. 7; Craig, *Psychological Medicine*, p. 415. The historian Benoît Majerus uses the term "diagnostic circularity" to sum up how doctors viewed patients' narratives: both as a sign of illness and as being discredited by illness. Benoît Majerus, *Parmi les Fous: Une Histoire Sociale de la Psychiatrie au XXe Siècle*, Rennes: Presses Universitaires de Rennes, 2019, p. 24.

awareness of themselves and their condition, their attitude to hospitalisation and to their relatives, and their relationships with staff and other patients in the institutions.

A first remark is that the "present condition" section of the case histories recorded many and varied physical problems, usually fatigue, pain, insomnia, loss of appetite, fever, paralysis, ophthalmological problems, dysarthria, tics, dyspnoea, menstrual and sexual disorders. Beyond the physical symptoms, symptoms can be classified into emotional, cognitive and those related to the behaviour and personality of the patients. Emotional symptoms were primarily melancholy, anxiety, fear, despair, excessive cheerfulness and euphoria, mental agitation manifested by intense emotions and irritability, excessive or shallow affect, sudden and unexplained mood swings, uncontrollable laughing and crying, and the absence of "normal" emotions such as love for family or sadness in the face of personal, family or social adversity.

Cognitive symptoms included cognitive impairment and slowness, foolishness, blunted or absent judgement, orientation and consciousness, disjointed or incoherent speech and difficulty in communication. Many patients were found to suffer from hallucinations (for example, believing that whatever they heard was referring to them) and delusions (for example, seeing demons, animals and people, hearing voices and feeling insects on their bodies). Inconsistent, foolish or bizarre ideas were also often recorded. These were usually categorised as delusions, mainly of persecution (when patients believed that they were being watched, threatened, bewitched or poisoned) and grandiosity (when patients believed that they had great power and wealth, and, in the case of women, that they were the most beautiful and that their household and family depended entirely on them). Conversely, others expressed ideas of failure, despair, guilt, death and doom, often associated with loss of desire for life, suicidal tendencies and hypochondriacal ideas. In some cases, delusions of jealousy or of religious content were recorded, with patients talking about the Second Coming or believing that they were possessed by the devil or that they were prophets.

Finally, the "present condition" described the patients' behaviour and character. Behavioural symptoms were often recorded first, as they formed the doctors' first impression of the patients: apathy, stillness, muteness and isolation; resistance to admission; agitation (continuous, purposeless and restless movements, verbosity, singing, dancing, laughing and shouting, constant hyperactivity, causing damage and disorder); aggression and violence; bizarre behaviour (for example, repeating the same phrases or movements and lying on the floor). In terms of character, patients were described as egocentric, quarrelsome, stubborn, immoral and unable to control themselves. Some patients were considered to be indecent and offensive, malicious, contemptuous, hypocritical, liars, irresponsible, unreliable, unwilling or unable to work and to take care of themselves. Exclusively women were described as flirting, exposing themselves and gazing seductively. Apart from the tendency to eroticism, other female symptoms were, as already noted, grandiose ideas about beauty and role in the family, but also the inability or unwillingness to perform household duties. Male pathologies, more than female ones, were characterised by delusions of power and wealth and an inability to engage in intellectual work, a symptom more common among middle- and upper-class men. Again, the identification and description of symptoms were clearly influenced by the social characteristics and roles attributed to the patients according to their gender and social class.

The analysis of diagnoses and symptoms leads to conclusions regarding psychiatric theory and practice, as well as the social use of psychiatric hospitals in Greece. It highlights the problems and ambiguities of psychiatric assessment, especially in cases whose characteristics can be interpreted not as thought, mood or personality disorders, but as reactions to the psychiatric examination and hospitalisation. Both diagnostic theory and practice were interpretive processes, influenced by the scientific and ideological orientations of physicians and the social and cultural perceptions and concerns of their time. The criteria for identifying the pathological were largely shaped by what was prescribed as normal according to gender, age and social class, as evidenced by the different frequency and meanings of diagnoses in men and women, children, adolescents, adults, and the elderly, and the higher and lower social classes.

Apart from social origins, diagnoses and symptoms also had social repercussions, directly or indirectly determining which lifestyles and behaviours were deviant, unethical, anti-social or dangerous for the patient and the community and proposing their reform. Those psychiatric diagnoses that were broad, encompassing physical, mental, emotional and moral disorders, were best suited to the multifaceted and difficult-to-diagnose conditions presented to psychiatrists and enabled the reconciliation of psychiatrists' scientific authority—signified by the clinical and organically articulated language—and their claims to a public role—expressed through analyses of personality and character of their patients. Especially some symptoms, such as laziness, impulsivity and moral indifference or perversion, as well as specific diagnoses, such as moral insanity, degeneration, alcohol and substance abuse and addiction, hysteria and neurasthenia, had stronger social and moral overtones. They better served psychiatry's attempts to regulate social relationships, establish behavioural, sexual and ethical standards, take on a reformative role, and voice proposals for important social issues of their time, such as poverty, crime, urban life and women's emancipation. In some cases, psychiatrists even linked the physical, mental and moral standards they outlined with the general interests of the nation and race, presenting it as their duty to defend these standards and providing guidelines on lifestyle, work, entertainment, marriage and family.

However, the doctors at the Eginitio and the Dromokaitio assessed that most of their patients suffered from severe and intense physical, mental and emotional symptoms, which were only occasionally combined with moral and social disorders, such as apathy, impulsiveness and sexual perversion. Such disorders never monopolised the patients' clinical picture as noted in the present condition and conveyed in the diagnosis. This agrees with the observation of some psychiatrists that patients who did not have any noticeable physical disorders and who did not disturb their surroundings, especially with noisy and persistent delusions, were not admitted to psychiatric institutions.[176] In other words, conditions without severe physical and mental symptoms were not commonly perceived as illnesses requiring psychiatric hospitalisation. In conclusion, psychiatric practice in early twentieth-century Greece, still largely represented by large neuropsychiatric hospitals, did not reflect the opening of psychiatry to nervous and mild disorders nor did it echo or fulfil the claims of the psychiatric profession (in Greece and abroad) for a broad public role as a signifier and guardian of individual and national health, ethics and progress.

[176] Vlavianos, "Alcoholism"; Yanniris, "On diseases caused by alcohol in Greece".

PART III

Psychiatric Hospitalisation

So far, the relationship between the institutions and their social environment has been analysed at the level of the institutions' management, admission policies and funding (Part I). It has also been shown that the likelihood and purposes of hospitalisation were shaped by established notions of mental illness, psychiatric hospitals and social characteristics, mainly gender, age and class (Part II). The third part of the book focuses on psychiatric hospitalisation in terms of admission and discharge decisions. It examines the ways in which psychiatric institutions were approached and used by patients and their private and public milieus, asking questions about who decided when and why hospitalisation began and ended, how long hospitalisation lasted, what alternatives were available, and what functions the institutions performed for the medical profession.

CHAPTER 9

Private Aspects of Institutional Mental Health Care

The patients' private environment, including family, friends, neighbours and other community members, was an important player in institutional mental health care. It was mainly family members who paid the hospital fees (see Chapter 3), but also applied for admission, were responsible for the patients, brought them to the institution, provided information about them and requested their discharge. Although information on all these aspects is not equally available in the two institutions, there is sufficient evidence to show that the Eginitio and the Dromokaitio were used by the patients' private environment and sometimes by the patients themselves, and to illustrate the different ways in which private intervention in and management of hospitalisation took place.

Private Admissions

A key area of private involvement in psychiatric hospitalisation was the request for admission. In the Dromokaitio, most patients were admitted by private application, while in some cases a private applicant was combined with a public applicant, for example a relative with the municipal authorities (Table 9.1). Private applications tended to be made by men (over 80% of private applicants) and relatives (about 60% of private applicants had the same surname as the patient). Moreover, even for

D. Kritsotaki, *A Social History of Mental Health Care in Greece, c. 1900–1920*, Mental Health in Historical Perspective,
https://doi.org/10.1007/978-3-032-05764-8_9

Table 9.1 Private and public admissions to the Dromokaitio

Applicant	*Patients*	
	N	%
Private	768	79.9
Public (state and local authorities, various associations)	166	17.3
Private and public	22	2.3
Unknown	5	0.5
Total	961	100.0

Source: Sample from the registers, Dromokaitio, 1905–1909 and 1916–1920 (see Table 1.1).

patients with a different surname from the applicants, we cannot exclude family relationships, not only because relatives do not always have the same surname, but also because non-relatives could act "on behalf of the family".[1]

Looking more closely at the private applications made by individuals with a stated relationship to the patients, the role of men and close relatives (siblings, parents, spouses and, less frequently, adult children) stands out. Applications by non-relatives were rare, as in the case of a 17-year-old unmarried "worker, farmer", who was admitted at the request of his "director".[2] In some cases, a lawyer, doctor or priest was recorded as the applicant, but this may have been because the relatives were unable to write the application themselves.[3] In any case, it is evident that the admission process was mainly carried out by the family, especially its male members, although this does not mean that it was predominantly male relatives who identified the problem and decided to have the patient admitted, nor that hospitalisation was decided unanimously. Unfortunately, the sources only rarely provide information about who detected

[1] *Βιβλίον των κατ' Αίτησιν Ιδιωτών Εισαγομένων Φρενοβλαβών* (*Book of the Insane Admitted at the Request of Private Individuals*, henceforth Dr/Register) 1918–1919, year 1919, no 60, archive of the Dromokaitio Hospital, Athens.

[2] Dr/Register, 1904–1906, year 1905, no 70.

[3] Dr/Register, 1906–1909, year 1907, no 26; 1915–1917, year 1916, no 40; 1919–1921, year 1920, no 66.

the problems and about conflicting family opinions about the patient's condition and the appropriate way to deal with it.[4]

In the Eginitio, the registers did not record the person requesting admission, but rather the patients' "relatives or guardians", and this occasionally, in only 122 cases. These "relatives or guardians" were mainly men, and in 32 cases they had the same surname as the patients, but the exact relationship was not specified, except in two cases where it was stated that they were the patients' sons.[5] More information is provided by the applications for free or reduced-fee hospitalisation, many of which were made by the patients' families, usually the father, mother or spouse, but also other relatives such as adult children or siblings, as well as the patients' guardians and sometimes the patients themselves. Occasionally, non-relatives from the immediate environment of the patients intervened and requested some form of preferential treatment: for example, a monk's monastery[6] and a student's school.[7]

The involvement of the patients' private environment is evidenced in two other admission procedures that did not necessarily coincide with the application for admission: bringing the patients to the institution and informing the doctors about their medical history. On the basis of the cases in which this information was recorded, it can be concluded that patients were usually led to the institution by relatives[8] and, to a lesser extent, by colleagues[9] or their doctor.[10] Members of the patients' private environment also played an important role in providing the patients' history. In the Eginitio, this was most often done by close relatives of

[4] See in the third section of the chapter for a case of disagreement between siblings about the admission of a patient to the Dromokaitio.

[5] *Γ' Μητρώον Εσωτερικών Ασθενών Αιγινήτειου Νοσοκομείου* [*Third Register of Inpatients of the Eginitio Hospital*], archive of the Eginitio Hospital, Eg/1.1, Athens (henceforth Eg/1.1), nos 2295 and 2413.

[6] Letter dated 27/4/1909, *Κλινικές-Νοσοκομεία* [*Clinics-Hospitals*], 1908–1909, Archive of the Rectorate of the University of Athens, Historical Archive of the University of Athens, Athens.

[7] Nevertheless, the fees were paid by the student's father. *Clinics-Hospitals*, 1918–1919, 12/7/1919.

[8] *Πρόχειραι Ιστορίαι Νόσου Εισαγομένων Ασθενών Β'* [*Draft Illness Histories of Patients Admitted B*], archive of the Eginitio Hospital, 9.2, Athens, p. 421 (henceforth Eg/9.2).

[9] Eg/9.2, p. 331.

[10] Eg/9.2, p. 377.

the patients, usually the mother and less often the father, husband or wife. Relatives providing information were most frequently women, while non-relatives (friends, neighbours, colleagues or doctors) were always men. In the Dromokaitio, the patients themselves and their relatives were usually mentioned as informants, but the exact relationship was not recorded except in two cases: the employer of one patient and the brother of another.[11]

In brief, the patients' private environment, and especially the family, was deeply and diversely involved in the initiation and early stages of hospitalisation. The central role of the patients' close relatives in the application process is related to the legal status of the relatives of those considered to be of diminished responsibility, and at the same time confirms the observation of doctors of the time that families sought to keep mental illness from being known outside the immediate family circle.[12] In addition to applying for admission, the private milieu of the patients assumed responsibility for them, led them to the institution and provided information about them. These procedures appear divided between men and women. Men were more likely to apply for admission and to take responsibility for the patients, as these were public procedures and thus closer to men's social roles. Women had the main role in giving the patient's history, a process that was more related to the intimate relationship with the patient. However, the division of these admission tasks between men and women was not absolute; the person making the application was often also the informant.

Motives and Purposes of Private Admissions

Why did the patients' private environment turn to institutions? Historians have shown that families used institutions to care for, treat, control, and reform sick and difficult members.[13] Some researchers, in fact, subscribe

[11] *Βιβλίον Ιστορικών [Case Histories Book]*, archive of the Dromokaitio Hospital, Athens (henceforth Dr/Case histories), 1916–1917, p. 773; 1918–1919, p. 998.

[12] Patricia Prestwich, "Family strategies and medical power: 'Voluntary' committal in a Parisian asylum, 1876–1914", *Journal of Social History,* 1994 (27), pp. 799–814.

[13] Mark Finnane, "Asylums, families and the state", *History Workshop Journal*, 20 (1985), pp. 134–48; John Walton, "Casting out and bringing back in Victorian England: Pauper lunatics, 1840–70", in William Frederick Bynum, Roy Porter and Michael Shepherd (eds), *The Anatomy of Madness: Essays in the History of Psychiatry,* v. 2, London:

to the view that by the nineteenth century family tolerance had waned, and destabilising members were more likely to be removed from the home. This change has been attributed both to the emergence of family ideals of harmony and stability since the mid-eighteenth century and to the increasing availability of institutions.[14] In this section, I will use the case histories from the Eginitio and the Dromokaitio, and in particular the section "history of illness", which details the condition of patients prior to admission, to understand what drove the private environment into the decision of hospitalisation.

First of all, it should be noted that the admission of many patients to the institutions was related to the severity of their physical condition; Sophia P., for example, was admitted to the Eginitio in a "subcomatose condition" with general weakness and loss of strength.[15] Physical symptoms were identified on admission not only by the doctors (see Chapter 8), but also by their relatives, who highlighted such symptoms as significant reasons for hospitalisation. For example, Maria K. had mental symptoms for a year and a half but was admitted to the Dromokaitio only when she couldn't get out of bed and had "urinary

Tavistock, 1985, pp. 132–46; Constance McGovern, "The community, the hospital, and the working-class patient: The multiple uses of asylum in nineteenth century America", *Pennsylvania History*, 54 (1987), pp. 17–33; David Wright, "Getting out of the asylum: Understanding the confinement of the insane in the nineteenth century", *Social History of Medicine*, 10 (1997), pp. 137–55; Patricia Prestwich, "Family strategies and medical power: 'Voluntary' committal in a Parisian asylum, 1876–1914", in Roy Porter and David Wright (eds), *The Confinement of the Insane: International Perspectives, 1800–1965*, Cambridge: Cambridge University Press, 2003, pp. 79–99; Yannick Ripa, *Women and Madness: The Incarceration of Women in Nineteenth-Century France*, Cambridge: Polity Press, 1990, ch. 2; Louise Wannell, "Patients' relatives and psychiatric doctors: Letter writing in the York Retreat, 1875–1910", *Social History of Medicine* 20 (2007), pp. 297–313; Catharine Coleborne, *Madness in the Family: Insanity and Institutions in the Australasian Colonial World 1860–1914,* New York: Palgrave Macmillan, 2010, p. 3; Anatole Le Bras, *Aliénés—Une Histoire Sociale de la Folie au XIXe Siècle*, Paris: CNRS Éditions, 2024, ch. 2 and 3.

[14] Nancy Tomes, *The Art of Asylum-Keeping: Thomas Story Kirkbride and the Origins of American Psychiatry*, Philadelphia: University of Pennsylvania Press, 1994 (first published in 1984), p. 127; Edward Shorter, *A History of Psychiatry: From the Era of the Asylum to the Age of Prozac*, New York: John Wiley, 1997, pp. 50–52; Roy Porter, "Madness and the family before Freud: The view of the mad-doctors", *Journal of Family History,* 23 (1998), pp. 159–72.

[15] Eg/9.2, p. 313.

and faecal incontinence".[16] Patients with critical physical problems, such as paralysis and inability to eat, were difficult to keep at home because they needed intensive and sometimes specialised care.

The difficulty of home care was not only due to physical disorders, but also due to the behaviour of patients who, according to their relatives, were unmanageable or intolerable and thus a strain to those around them.[17] In particular, suicide threats and attempts caused concern and contributed to patients being placed in more closely supervised environments.[18] Characteristically, patients who had not been well for years were admitted within days or weeks after a suicide attempt.[19] Other cases involved apathy and unwillingness to care of themselves.[20] One patient's "absentmindedness", his brother noted, required his constant and careful supervision.[21] When this kind of care and supervision was not possible, the patients' environment turned to the institutional solution, especially in cases where patients became hostile or violent towards those who cared for them. Agitation, violence, and destructiveness were among the behaviours most frequently cited by patients' relatives as reasons for admission, signalling the escalation of tension in family relationships. For example, Christos L. was transferred to Athens for treatment by his relatives only when he began to use obscene language against them; Yeoryios T. "harboured a deep hatred for his father and sister, so he was admitted

[16] Dr/Case histories, 1904–1907, p. 811.

[17] On patient behaviour as a factor in psychiatric admissions, see Joost Vijselaar, "Out and in: The family and the asylum. Patterns of admission and discharge in three Dutch psychiatric hospitals 1890–1590", in Marijke Gijswijt-Hofstra, Harry Oosterhuis, Joost Vijselaar and Hugh Freeman (eds), *Psychiatric Cultures Compared: Psychiatry and Mental Health Care in the Twentieth Century: Comparisons and Approaches*, Amsterdam: Amsterdam University Press, 2006, pp. 277–94. Families were particularly struggling to care for elderly members with serious illnesses, such as dementia, which combined intense physical symptoms with challenging behaviour. Anthea Vreugdenhil, "'Incoherent and violent if crossed': The admission of older people to the New Norfolk Lunatic Asylum in the nineteenth century", *Health and History*, 14 (2012), pp. 91–111.

[18] *Πρόχειραι Ιστορίαι Νοσούντων Α΄* [*Draft Histories of The Ill A*], archive of the Eginitio Hospital, 9.1, Athens, p. 7 (henceforth Eg/9.1).

[19] Dr/Case histories. 1918, pp. 950–51.

[20] Eg/9.2, pp. 147, 193.

[21] Letter of the patient's brother in Dr/Case histories, 1900–1902.

to the hospital again", and Iakovos R., "an alcoholic who became aggressive ten days ago, was admitted to the clinic".[22] The patients' behaviour could be seen as disruptive beyond their home, to their workplace or neighbourhood.[23]

The definition of uncontrollable behaviour was related to gender and age. Relatives of male patients associated illness and the need for hospitalisation with unemployment and especially an unwillingness to work.[24] Conversely, in the case of women, families stressed the neglect of domestic duties, which was sometimes linked to alcohol consumption and violent behaviour. For example, one patient's condition worsened when she began to drink "excessively". She started wandering around, going to houses and shops "indecently", and by the time she was admitted she could no longer take care of her household. Her thinking was clear, and she was calm on admission. Her only symptoms were vomiting, cessation of menstruation and emotionality. Therefore, the reasons for her admission centred on the violation of predefined female roles and characteristics: taking care of the household, staying in the private sphere and showing restraint and modesty.[25] With regard to age, the relatives of some children and adolescents emphasised that they did not work, made unreasonable demands or did not behave in a manner appropriate for their age and as advised by their elders, even in simple matters such as washing their hands and making the sign of the cross.[26]

Many of the behaviours mentioned so far were associated with immorality by the patient's family, particularly when the diagnosis was "moral insanity". The cases of two adolescents, Charilaos A., 18 years old, and Triantafyllos T., 17 years old, are emblematic. The former, described as a capricious youngster and a thief since childhood, lied to avoid school and to extort money from people he knew, and left home without reason or notice. The medical assessment agreed with the descriptions of his family members: according to the doctor at the Dromokaitio, Charilaos had "mental abnormalities in the moral and intellectual sphere" such as

[22] Respectively: letter by the patient's brother in Dr/Case histories, 1900–1902; letter by a patient's relative in Dr/Case histories, 1900–1902; Eg/9.2, p. 80.

[23] Eg/9.1, p. 187.

[24] Dr/Case histories, 1918, pp. 920–21.

[25] Dr/Case histories, 1907–1912, p. 4.

[26] Eg/9.1, p. 135.

apathy, abulia and "rapid extinction of affectionate feelings", and was diagnosed with "degenerative or hereditary insanity, moral insanity with periodic agitations".[27]

A similar diagnosis ("hereditary or degenerative insanity in the form of moral insanity") was given to Triantafyllos T., whose "moral perversions" and uncompromising character had also manifested themselves since childhood, causing various episodes in the family. He was strong-willed, irritable, impulsive and "not to be deterred by anything". He did not want to go to school and was driven there by force or even by police intervention. His demand for money led to intense family conflicts, during which he blackmailed his parents and prevailed upon them, extorting money from them to satiate his "lavish" desires, which involved "common women" and friends of lower social class. When his parents resisted or tried to admonish him, he would become angry, use obscene language, assault them and resort to "criminal acts". The inability of his parents to control and correct him, the daily deterioration of his mental state "as a consequence of the reprimands of his relatives", and his good intellectual state on admission—the doctor noted that he did not show any significant symptoms—all reinforce the impression that Triantafyllos was admitted to the Dromokaitio for disciplinary reasons.[28]

In sum, the evidence from the institutions confirms that in Greece, as in other countries, psychiatric hospitals were used by the private environment of patients as a way out of a usually intense and sometimes prolonged crisis. Patients were mainly admitted when they were causing problems, agitation and disturbance; when they were considered a threat to others or to themselves; and/or when they needed special treatment and care.[29] The purpose of hospitalisation was not only to cure and care for patients, but also to relieve the family of the burden and responsibility of caring for and supervising difficult members. The example of a university student whose mother requested his hospitalisation in the Eginitio because he was endangering his life and the lives of everyone at home, and it was impossible to care for and supervise him is typical. The mother stressed that she would not object to her son being admitted

[27] Dr/Case histories, 1920, p. 1271.

[28] Dr/Case histories, 1917, pp. 872–73.

[29] Tomes, *The Art of Asylum-Keeping*, pp. 108–13; Prestwich, "Family strategies and medical power"; Vijselaar, "Out and in: The family and the asylum".

at the university's expense to the Dromokaitio, where he could be cared for "for life".[30] The institutions were therefore called upon to take on long-term care and a custodial role.

At the same time, the association of psychiatric hospitalisation with out-of-control behaviour, perceived as immoral, dangerous, and incompatible with the roles and characteristics expected of patients, suggests that psychiatric institutions could be used by the private environment, and especially the family, as a means of discipline and reform.[31] In these cases, the family appears not as the private space of protection described by the domestic ideal, but as an agent of social control, sometimes using psychiatry to achieve this.[32] Could we conclude, then, that the efforts of psychiatrists to play a role in the management of social relations outlined in Part II (see mainly Chapter 8) had had an impact?

The answer is only partly affirmative. Uncompromising and uncontrollable behaviour was described in almost all patients, making it one of the main features of mental illness as perceived by doctors and non-doctors alike. Most often, however, it was part of a wider picture of disorders such as incoherence, suicidal tendencies, intellectual impairment,[33] memory problems, depression, and delirium.[34] In short, although the institutions admitted individuals who looked less like mentally ill and more like difficult or different people, the condition of most patients was perceived and described by their relatives as serious and critical. As concluded in Chapter 8, conditions limited to behavioural problems without severe physical and mental symptoms were not commonly perceived as requiring psychiatric hospitalisation.

Nevertheless, in the nineteenth and early twentieth centuries, the confinement of healthy individuals dominated public perceptions of the psychiatric institution. In Britain, for example, autobiographical writings by former inmates, newspaper articles and literary works revolved around

[30] Letter dated 7/10/1915, *Clinics-Hospitals*, 1915–1916, Archive of the University of Athens.

[31] Ripa, *Women and Madness*, ch. 2.

[32] Elizabeth Lunbeck, *The Psychiatric Persuasion: Knowledge, Gender, and Power in Modern America*, Princeton, N.J.: Princeton University Press, 1994, p. 96.

[33] Dr/Case histories, 1907–1912, p. 134.

[34] Dr/Case histories, 1917, pp. 872–73.

and fuelled debates about wrongful confinement—the psychiatric hospitalisation of mentally healthy people by their families to get rid of them, take their property or punish them.[35] Similar opinions and criticisms of the psychiatric system were expressed in France in the second half of the nineteenth century; as scandals of wrongful confinement emerged, the asylum was increasingly portrayed as an instrument of exercising family authority and resolving family conflicts.[36] In early twentieth-century Greece, similar cases were reported in the newspapers. For example, the newspaper *Εμπρός* [*Forward*] published a series of articles on a man who claimed that he had been admitted to the Dromokaitio by his brother for family reasons that remained unspecified. According to the newspaper's correspondent, the victim of the injustice was supported by public opinion,[37] which believed that "victims of personal revenge were being held there [in the Dromokaitio]".[38]

Regardless of the extent to which such articles reflected public opinion, incidents of incarceration for personal and financial reasons attracted public interest and often took on a public dimension. In psychiatric discourse, however, they were treated differently. On the one hand, psychiatrists complained that the legal procedures established in many countries, including Greece, to avoid wrongful confinement were not only unnecessary but also detrimental to patients' health because they delayed treatment.[39] On the other hand, psychiatrists usually insisted on the correctness of their diagnosis. In the case mentioned above, the director of the Dromokaitio insisted that the man was mentally ill, suffering from chronic systematised delusion.[40] In the Eginitio and the

[35] Joshua John Schwieso, "Religious fanaticism and wrongful confinement in Victorian England", *Social History of Medicine*, 9 (1996), pp. 159–74; Roy Porter, *Madness, A Brief History*, Oxford: Oxford University Press, 2002, pp. 167–72.

[36] Jessie Hewitt, *Institutionalizing Gender. Madness, the Family, and Psychiatric Power in Nineteenth-Century France*, Ithaca: Cornell University Press, 2020, pp. 15–16, 92.

[37] Anon., "Αι αποκαλύψεις του φρενοκομείου" ["The revelations of the lunatic asylum"], *Εμπρός* [*Forward*], 16/10/1901, pp. 1–2.

[38] Anon., "Τα του φρενοκομείου" ["Issues of the lunatic asylum"], *Εμπρός* [*Forward*], 18/10/1901, p. 1.

[39] For example, Thomas Clouston, *Clinical Lectures on Mental Diseases*, 5th edition, London: J. & A. Churchill, 1898, p. 671. On "wrongful confinement" in Britain, see Peter McCandless, "Liberty and lunacy: The Victorians and wrongful confinement", *Journal of Social History*, 11 (1978), pp. 366–86.

[40] Anon., "The revelations of the lunatic asylum".

Dromokaitio, doctors regularly perceived as symptoms of mental illness the occasional claims by patients that they had been wrongly admitted by relatives for financial reasons, e.g. to seize their property, or personal reasons, e.g. to punish them.[41] Although it is impossible to say who was right in such cases, the doctors, the relatives, "public opinion" or the patients, it is crucial to highlight the difference between psychiatric and lay discourses. The former promoted the legitimate operation of the institutions, stressing that the reasons for admission had to do with the clinical condition of the patients, while the latter, formulated by patients, newspapers and novels, emphasised the scandalous use of psychiatric hospitalisation and the personal and family motives behind it.

Private Initiative Discharges

In Greece, as in other countries, the private environment of patients, and especially the family, played a central role not only at the beginning but also at the end of hospitalisation, demanding the discharge of private or even public patients, sometimes against the doctors' advice.[42] As mentioned in Chapter 3, according to the psychiatric law, patients were discharged by medical decision when they were cured or when a relative or the person who had applied for admission requested their discharge. In the Eginitio and the Dromokaitio, the majority of patients were discharged not cured but improved or in the same condition (Table 9.2). A major reason for discontinuing hospitalisation, especially when patients paid out of their pocket, was the inability to cover the fees (see Chapter 4). In addition, it can be assumed that financial reasons motivated private decisions to transfer a small number of patients (only four were recorded) from the Dromokaitio to the public psychiatric hospitals of Corfu and Athens, where hospitalisation was free.

As a rule, though, not much information is provided on discharges. In the Eginitio, it is not recorded who requested the discharge. In

[41] Dr/Case histories, 1918, pp. 914–15; 1907–1912, p. 4.

[42] On the key role of families in hospitalisation and the dynamic relationships and negotiations of family members with the medical and lay administration of institutions and public authorities, see Anna Shepherd, *Institutionalizing the Insane in Nineteenth-Century England*, London: Pickering & Chatto, 2014, ch. 3; Samuel Dal Zilio, "'Réclamée par sa famille contre l'avis des médecins': Pouvoir et implication des familles de patients dans les processus d'internement psychiatrique à l'asile de Schaerbeek entre 1850 et 1914", *Journal of Belgian History*, 51 (2021), pp. 100–24.

Table 9.2 Hospitalisation outcomes

Outcome	*Patients admitted to the Eginitio*		*Patients admitted to the Dromokaitio*	
	N	%	*N*	%
Cure	277	19.4	121	12.6
Improvement	537	37.6	108	11.2
No improvement/stable	445	31.1	415	43.2
Death	151	10.6	286	29.8
Unknown	20	1.4	31	3.2
Total	1430	100.0	961	100.0

Source: Sample from the registers, Eginitio and Dromokaitio, 1905–1909 and 1916–1920 (see Table 1.1).

the Dromokaitio registers, this was recorded for 374 patients who were discharged without being cured. About 75% of these patients—almost all of whom had been admitted on private request, but also a small number of public patients—were discharged on private request, usually by the mother, father, brother, husband or wife. Women appear to have played a greater role in discharging than in admitting family members, a trend noted by other researchers and considered consistent with women's role in the family as carers.[43] Indeed, in some cases the mother requested the discharge of a patient who had been admitted at the request of the father.[44] However, the reverse also occurred[45] and it would thus be simplistic to distinguish admission and discharge as a male or female matter. Besides, discharge to the family did not always mean that it was the family who initially requested the discharge: doctors could also recommend discharge of unimproved patients, and their families were more or less obliged to accept them. Moreover, the family's decision to request discharge, like the decision to request admission, was not always unanimous.[46] For example, in the above-mentioned case of "wrongful confinement" in the Dromokaitio, the patient's brother had stipulated the admission, while their sister opposed to it. During a visit to the institution,

[43] Ripa, *Women and Madness*, pp. 39–42.

[44] Dr/Register, 1919–1921, year 1920, nos 23 and 19.

[45] Dr/Register, 1919–1921, year 1920, no 55.

[46] Tomes, *The Art of Asylum-Keeping*, p. 121.

she helped the patient to escape.[47] This story of a patient's eventful recuperation illustrates the family tensions and legal obstacles that sometimes complicated the private uses of psychiatric hospitalisation.

One distinct type of discharge, common in Western psychiatric institutions, was the trial discharge: the patients left the institution, but their admission documents remained valid, allowing for easy and quick readmission should the need arose.[48] In the Dromokaitio, most of the patients who were discharged to their families but soon returned to the institution had been discharged on a trial basis. Some were readmitted more than once over a long period, such as Ismini P., who was first admitted at the age of 19 and who was received by her family on a trial basis and readmitted eight times in six years.[49] Although it is not possible to ascertain whether these discharges were initiated by the family or by the hospital authorities, the trial discharges and readmissions show that the institutions were not places of permanent and irreversible isolation, but had permeable boundaries, and could to some extent be used by the patients' private environment according to its possibilities and needs on the one hand, and the patients' condition on the other.

The Patients' Role

Patients cannot be missing from the discussion of the private aspects of psychiatric hospitalisation. But while patients were always present and important informants for doctors—either through their narratives or through their own bodies, speech and behaviour—they were largely absent from hospitals' records because as "lunatics" they were largely considered irresponsible and incapable of making decisions for themselves. Indicatively, their consent was not required for admission, and they could not discharge themselves. In some cases, however, it is possible to speculate on how they viewed their hospitalisation based on how their condition and symptoms were recorded by doctors on admission and during their stay in the institutions.

47 Anon., "Ο δραπέτης του Δρομοκαΐτείου" ["The escapee of the Dromokaitio"], *Εμπρός* [*Forward*], 15/10/1901, p. 2; Anon., "The revelations of the lunatic asylum".

48 Muriel Bonaïti, "Histoire de l'asile XIXe siècle: D'un lieu unique de soin aux premières alternatives à l' hospitalisation", PhD thesis, Université de Lyon, 2007, p. 70; Vijselaar, "Out and in: The family and the asylum".

49 Dr/Register, 1917–1918, 1918, no 24.

In this way, we know that many patients disagreed with their hospitalisation, either because they thought it was unnecessary, because they believed they were incurable, or, more often, because they did not believe they were mentally ill at all, describing their admission as illegal and demanding to see the necessary certificates.[50] As we have seen, some patients believed that they had been wrongly admitted, either to serve their interests of their relatives or to be punished.[51] For instance, the soldier Nikolaos K. claimed that he had been admitted to the Dromokaito to be reprimanded by his superiors.[52] Triantafyllos T., who, as mentioned above, was admitted mainly because his parents could not control him, saw his admission as a punishment: he said that he had been punished enough for what he had done, admitted that he had made a mistake, and promised that he would obey his parents in future.[53] Many patients complained about their detention, turned against the doctors, threatening to prosecute them or were enraged with their relatives who had brought them there.[54] Charalambos D., when he realised that his brother was taking him to the Dromokaitio, became angry and cried because he would be stuck with "the title of madman", but concluded that he had to give in "before force majeure".[55] Moreover, many were eager to be released, like Triantafyllos T., who was asking the doctors: "What do you think of me, am I still not well?"[56] Others asked to leave, either with threats and anger—like Olvieros L., who in an "angry" and aggressive tone asked for the door to be opened so that he could leave—or with tears and pleas, like Maria B., 15 years old, who cried and nodded, begging to be sent to her relatives, because it was impossible for her to be "among the mad".[57]

However, patients could not leave on their own accord, and it seems that they had limited ability to persuade doctors and relatives to allow

[50] Eg/9.2, pp. 107, 113; Dr/Case histories, 1907–1912, p. 118; 1916–1917, pp. 773–74.

[51] Dr/Case histories, 1918, pp. 914–15; 1907–1912, p. 4.

[52] Dr/Case histories, 1920, pp. 1346–47.

[53] Dr/Case histories, 1917, pp. 872–73.

[54] Dr/Case histories, 1907–1912, p. 4; Eg/9.2, pp. 331, 1.

[55] Dr/Case histories, 1918, pp. 938–39.

[56] Dr/Case histories, 1917, pp. 872–73.

[57] Dr/Case histories, 1916–1917, pp. 773–74; 1907–1912, p. 118; 1918, pp. 926–27.

them to leave. In this context, some attempted to escape. At the beginning of the twentieth century at least three patients attempted to escape from the Eginitio and six managed to escape from the Dromokaitio. The family could, and sometimes did, keep escapees who returned home.[58] Other escapees were returned to the institutions by the police. Also, patients' reluctance and opposition to hospitalisation may be behind some of the suicide attempts in institutions[59] and may explain some of the patients' attitudes recorded in the case histories, such as refusal to talk, eat or cooperate,[60] as well as acts of violence: for example, Ioannis L. injured a male nurse on the grounds that he was harassing him, and soldier Nicholas K. tried to abuse those giving him a bath.[61] Whatever form resistance took, it was usually included by doctors in the symptoms of the disease. Quiet versions of resistance—such as apathy and isolation—were considered part of the clinical picture of melancholia and catatonia, in which the symptom of negativism in particular described the patients' resistance to the dictates of their own needs and the environment (in this case, the hospital staff).[62] More aggressive forms of resistance were attributed to agitation or delusional ideas. While we cannot exclude the possibility that oppositional behaviour was indeed an aspect of the patients' condition, it is also possible that it was related to their hospitalisation against their will. In any case, the institutions dealt with these behaviours using the psychiatric methods of the time, such as close observation in smaller wards, isolation, bed restraint, sedation and spoon or tube feeding.

Although complaints and intense, sometimes violent, behaviour were perhaps the most striking responses to hospitalisation, they were neither the only nor the most common. Some patients were quiet and indifferent to their new surroundings, without expressing a desire to leave.[63] Some

[58] Dr/Case histories, 1920, p. 1305.

[59] There were only a few recorded suicide attempts in the institutions: two (by hanging) in the Eginitio and none in the Dromokaitio. Two more suicides in the Eginitio were recorded in the archive of the University of Athens (*Clinics-Hospitals*, 1909–1910, 20/5/1910; 1911–1912, 23/4/1912).

[60] Dr/Case histories, 1904–1907, p. 626; 1907–1912, p. 118; Eg/9.1, pp. 139, 351.

[61] Dr/Case histories, 1918, pp. 914–15; Eg/9.1, p. 139.

[62] Dr/Case histories, 1904–1907, p. 626; 1916–1917, p. 785; Eg/9.1, p. 139.

[63] Dr/Case histories, 1907–1912, p. 27; 1916–1917, p. 808; 1918, pp. 950–51; 1907–1912, p. 72; Eg/9.2, pp. 139, 417, 355.

among them, however, did not understand where they were, thinking, for example, that they had gone for physical treatment, to visit someone, to spend some time in the country or even to get married.[64] Usually, such misconceptions were recorded in patients whom the doctors assessed as suffering from mental weakness[65] or delusions, as in the case of Emmanuel X., who believed that he was a millionaire, that he was going to build ships and that he was at the Eginitio to receive a captain's degree from Professor Katsaras and the King.[66] Although it is not documented in the patient records, it seems that sometimes patients were deceived into going to the hospital by those who admitted them.[67] For example, in the case of "wrongful confinement" mentioned above, the patient stated after leaving the Dromokaitio that he had been admitted there under false pretences: his brother had taken him to Athens ostensibly to meet the King, and then to the Dromokaitio supposedly to visit the patients.[68]

Other patients seemed pleased with their admission, but some of them were unaware of their surroundings and situation, or were described as "of limited intellect",[69] while others were satisfied for specific reasons. For example, Ageliki K., who was "not distinguished by her cleanliness", once admitted and bathed, was delighted and thanked the nurse.[70] Finally, some admitted that they were "crazy"[71] and were willing to cooperate with the doctors, such as Efrosini R., who described in detail the "disease" that "torments me night and day and makes me suffer all sorts of things, without having a moment's peace".[72] A few patients were admitted to the Eginitio on their own initiative, such as Dimitrios Ch., who "came to us requesting our scientific assistance" because of chronic vomiting,

[64] Dr/Case histories, 1904–1907, p. 824; 1907–1912, p. 39; 1918, pp. 896–97; 1907–1912, p. 24.

[65] Dr/Case histories, 1907–1912, p. 188; 1904–1907, p. 789; Eg/9.2, pp. 141, 277.

[66] Eg/9.2, p. 1.

[67] On the practice of deceiving patients into going to the psychiatric hospital, see Allan Beveridge, "Life in the asylum: Patients' letters from Morningside, 1873–1908", *History of Psychiatry*, 9 (1998), pp. 431–68.

[68] Anon., "The escapee of the Dromokaitio".

[69] Dr/Case histories, 1916–1917, p. 818; 1907–1912, p. 150; 1920–1921, p. 1351.

[70] Eg/9.2, p. 39.

[71] Eg/9.2, p. 117.

[72] Dr/Case histories, 1907–1912, p. 113.

headaches and weakness.[73] These patients had mostly physical symptoms, and their admission was connected to the particular physiognomy of the Eginitio as a university hospital with a neurological specialisation, located near the city centre. This type of admission, although limited, corresponds to an informal version of voluntary hospitalisation, a practice that already existed in some European countries.[74] As such cases demonstrate, the hospital could sometimes function as a place of treatment and care chosen by the patients, as has been shown in other countries, especially for patients who, before their admission, had lived in harsh conditions, in isolation and confinement or in a violent environment.[75] However, the general picture from both institutions is that patients rarely consented to their admission and even more rarely had a say in the decision to be hospitalised.

Conclusion

Voluntary or (usually) not, the private use of institutional mental health care was subject to the dynamics of the relationships between patients, their private environment, and their doctors; the balance between family members; the financial situation of patients and their relatives; the alternative care options available to them; as well as the legal and medical terms of psychiatric hospitalisation. The tensions between these parameters are evident in the case histories, which generally favoured the medical discourse and the accounts of the "healthy" (those responsible for the patients) over the lay discourse and the accounts of the "ill". Although patients had some direct or indirect control over their admission and discharge, usually their will was not the determining factor in their hospitalisation. However, institutions were by no means places of

[73] Eg/9.1, p. 101. See also Eg/9.2, p. 419.

[74] Gayle Davis, *"The Cruel Madness of Love": Sex, Syphilis and Psychiatry in Scotland, 1880–1930*, Amsterdam: Rodopi, 2008.

[75] Walton, "Casting out and bringing back"; Finnane, "Asylums, families and the state"; Janice Chesters, "A horror of the asylum or of the home: Women's stories 1880–1910", in Catherine Coleborne and Dolly MacKinnon (eds), *Madness in Australia: Histories, Heritage and the Asylum*, St. Lucia: University of Queensland Press, 2003, pp. 135–44; Tanja Luckins, "Crazed with grief: The asylum and the Great War in Australia", in Coleborne and MacKinnon (eds), *Madness in Australia*, pp. 169–79; Prestwich, "Family strategies and medical power".

absolute medical sovereignty; the processes surrounding the initiation and termination of institutional treatment were based on decisions made by the patient's private environment and especially the family, who negotiated the use of institutions according to their needs and the patient's condition.

In conclusion, in the early twentieth century, psychiatric institutions were in communication and interaction with their social environment. They were integrated into private strategies for treating and supervising serious physical and mental conditions, for dealing with problems in daily life and relationships, and for controlling difficult behaviour. In some respects, lay and psychiatric definitions of these conditions, problems and behaviours diverged, as the example of "wrongful confinement" illustrates: the inappropriate use of institutions by the family and the state was emphasised in lay and public discourses, but denied by psychiatrists. However, in many cases lay and psychiatric discourses overlapped, highlighting the continuities between psychiatric and lay uses of hospitalisation and the infiltration of medical theories and methods into lay understandings and practices.

CHAPTER 10

Public Aspects of Institutional Mental Health Care

Moving away from the private use of psychiatric institutions, this chapter examines the involvement of public agents in institutional mental health care in ways that go beyond the supervision and funding of hospitalisation (see Chapters 3 and 4). State and local authorities, charitable organisations, as well as the administrative and medical authorities of the institutions can be considered as public agents, in the sense that they were outside the private environment of patients and sought to admit or discharge patients for reasons that were usually related to the public manifestations of mental illness.

Public Admissions

In the sample from the Dromokaitio registers, about 17% of patients were admitted at the request of a public agent (Table 9.1), while the hospital's annual reports indicate that at the beginning of the twentieth century a larger proportion of patients, and in some years the majority, were hospitalised at public expense (see Chapter 4). Although it is possible that the public agents involved in hospitalisation were not always recorded in the registers, it seems that the public played a greater role in paying for hospitalisation than in initiating it, a task that was more often performed by the private environment of the patients (see Chapter 9).

D. Kritsotaki, *A Social History of Mental Health Care in Greece, c. 1900–1920*, Mental Health in Historical Perspective,
https://doi.org/10.1007/978-3-032-05764-8_10

Certainly, the regulation of the Dromokaitio stipulated that the institution could be used by the public in accordance with law 742 "On the establishment of lunatic asylums".[1] The registers recorded a greater variety of public agents than the law provided for.[2] The most important were not the police and the prefecture or municipality, as the law stipulated, but the Ministry of the Military, which was also a major funder of patients in the hospital. Thus, the main public use of the Dromokaitio was the hospitalisation of "psychopathic soldiers" and "psychopathic officers", who were usually admitted through military hospitals on the orders of the Ministry of the Military and with a simple medical indication, without medical certificates.[3] 143 of the 166 public patients in the sample were soldiers and officers, most of them admitted from 1918 to 1920, i.e. during the First World War and the Greco-Turkish War.[4] The landscape of public admissions to the Dromokaitio was completed by admissions requested by other ministries (Interior, Foreign Affairs, Navy, Health and Education), more occasionally by other public bodies such as consulates and embassies, and by the police and judicial authorities. Very few patients were admitted under the designation "dangerous": three men and two women by the police and five suspects by order of the public prosecutor "for observation" to determine whether they were mentally ill.

Unlike the Dromokaitio, the Eginitio, as already noted, was not subject to law 742 and was therefore not obliged to accept public patients. Certainly, the number of patients for whom an actor outside their private environment was noted in the registers was very small (39 or 2.7% of the patients admitted), but this information was not systematically recorded (see Chapter 9). An additional insight is provided by the letters kept in the archive of the University of Athens, in which public agents, often those we find in the registers, requested the admission of patients: ministries

[1] "Περί εγκρίσεως του προσωρινού Κανονισμού του συσταθέντος Φρενοκομείου του αοιδίμου Ζωρζή Δρομοκαΐτου" ["On the approval of the provisional Regulation of the established Lunatic Asylum of the Venerable Zorzis Dromokaitis"], *Εφημερίς της Κυβερνήσεως [Government Gazette]*, 192, 16/7/1887, article 72.

[2] Νόμος ΨΜΒ' "Περί συστάσεως φρενοκομείων" [Law 742 "On the establishment of lunatic asylums", *Εφημερίς της Κυβερνήσεως [Government Gazette]*, 28, 19/5/1862.

[3] Michael Yanniris, [Letter to the Prefect], 1918, *Α΄ Βιβλίο εξ Επαγγέλματος Εισαγομένων Ψυχοπαθών [First Book of ex Officio Admitted Psychopaths]*, archive of the Dromokaitio Hospital, p. 89.

[4] Indeed, only from 1918 did a separate register for public patients was kept in the Dromokaitio, as the law 742 (article 13) required.

(mainly the Ministry of the Military, but also the Ministry of the Navy, the Interior and Foreign Affairs), local authorities (prefectures, municipalities and communities) and only on one occasion the police.[5] These letters highlight the diversity and sometimes the combination of the public agents involved in the admission process. For example, the diocese of Athens requested that a patient sent from the diocese of Smyrna be treated on credit until the municipality of Athens approved to subsidise her hospitalisation.[6] In any case, while the public sought to send patients to the Eginitio, in practice the university hospital was less available for public use than the Dromokaitio. The latter, been subject to law 742, was the first and main option for public mental health care in early twentieth-century Greece, as evidenced by the fact that some public applications were made only after admission to the Dromokaitio had been rejected due to lack of beds.[7]

Regarding the small number of police admissions to both hospitals, this was to some extent related to the cost of hospitalisation and the shortage of beds. The main practice for the "dangerously insane" was to be detained in police stations or transferred to the public psychiatric hospital of Corfu (see Chapter 3), two options that were deemed both inappropriate and inadequate. According to an article in the newspaper *Εμπρός [Forward]*, the director of the police had informed the Ministry of the Interior that Athens was full of "lunatics" and that the five "most intimidating" of them were in police stations, unable to be transferred either to the Dromokaitio, because it was full, or to the Corfu hospital, which was under repair, or to their families, as they came from "abroad".[8] In the 1910s, an effort to address the problem was made by the establishment of the public psychiatric hospital of Athens, where admissions were mostly made by the police (see Chapter 3).[9] Patients admitted to

[5] *Κλινικές-Νοσοκομεία [Clinics-Hospitals]*, 1915–1916, 22/7/1916, Archive of the Rectorate of the University of Athens, Historical Archive of the University of Athens, Athens.

[6] *Clinics-Hospitals*, 1919–1920, 24/6/1920.

[7] *Clinics-Hospitals*, 1914–1915, 28/4/1915; 1913–1914, 22/8/1914.

[8] Anon., "Το Δρομοκαΐτειον γεμάτο!" ["The Dromokaitio full!"], *Εμπρός [Forward]*, 30/6/1902, p. 3.

[9] This was probably the same as the asylum for the mentally ill who were "dangerous to public safety" and "offensive to public decency", established in 1913. Ilias Makris (*Οι Φρενοπαθείς Εγκληματίαι Αλλαχού και εν Ελλάδι [The Insane Criminals in Greece and*

the Eginitio and the Dromokaio by the police would be transferred there if considered incurable of dangerous.[10]

Public involvement in admissions is not limited to admissions at public request. Even private applications sometimes involved non-private agents. Since hospital places were scarce and the system for administering free and low-cost hospitalisation was quite flexible, some private requests were sent through or supported by public agents, prominent individuals, members of "high society" and charitable associations, who acted as guarantors for the patients. For example, the request of Yeoryoula A. from Nafpaktos for her son to be treated free of charge in the Eginitio was forwarded by the Prefect of Aetolia-Acarnania in west-central Greece; the request of a father for his daughter to be hospitalised at low-cost was forwarded to the Eginitio by the Refugee Settlement Committee; and the admission request of Alexandros D. was addressed to the "Lady in Honour of the Queen's Majesty" and was forwarded to the Eginitio by the Patriotic League of Greek Women, a philanthropic association founded and headed by the Queen in 1914.[11] In the Dromokaitio, where there were more applications than could be met (see Chapter 4), the board of directors seriously considered the patient's guarantor or patron when deciding on both pauper and fee-paying patients. The intervention of guarantors extended to public admissions to the Dromokaitio, where, according to a later source, "bribery mechanisms" were at work and priority was given to those with the best "connections".[12] In addition, non-private bodies were involved in the hospitalisation through the assessment of applications for

Elsewhere], Athens: s.n., 1924, p. 28) a neuropsychiatrist specialising in forensic psychiatry, described the asylum in very negative terms, as inadequate, providing no treatment, and inhumane.

[10] *Clinics-Hospitals,* 1915–1916, 22/7/1916; *Βιβλίον των κατ' Αίτησιν Ιδιωτών Εισαγομένων Φρενοβλαβών [Book of the Insane Admitted at the Request of Private Individuals* (henceforth Dr/Register) 1917–1918, year 1917, no 19.

[11] *Clinics-Hospitals,* 1915–1916, 10/3/1916; 1914–1915, 28/4/1915; 1916–1917, 19/2/1917. On the Patriotic League of Greek Women, see Vassiliki Theodorou, "La Fondation patriotique pour la protection de l'enfant en Grèce. De l'action des philanthropes à l'organisation des services publics de santé et de prévoyance pour la maternité et l'enfance (1914–1940)", *Revue d'Histoire de la Protection Sociale*, 11:1 (2018), pp. 72–91.

[12] This was claimed by Michael Striggaris in 1949, cited in Lena Atzina, *Η Μακρά Εισαγωγή της Ψυχανάλυσης στην Ελλάδα [The Long Introduction of Psychoanalysis in Greece]*, Athens: Εξάντας [Exandas], 2004, p. 110.

low-cost and free hospitalisation: in the case of the Eginitio, the administrative and medical authorities, namely the rector, who was the public representative of the clinic and the first recipient of applications, and the medical director (see Chapter 12); in the case of the Dromokaitio, the board of directors.[13]

Finally, the patients' public environment was involved in the admission process by transporting them to the institution and providing information about them. Some patients were brought to the hospital by the police—some had been arrested before admission—because they were considered to be in a dangerous situation or had committed a crime. For such patients, their history was often provided by the police. In addition, at the Dromokaitio, official documents such as court records and reports from army officers and military doctors supplied information on patients awaiting trial and "service patients", respectively. However, escorting patients to the institution and providing information was more often undertaken by the patients' private environment.

Motives and Purposes of Public Admissions

As with private patients, public patients tended to suffer from severe physical symptoms. Bodily problems such as exhaustion, incontinence, and difficulty in eating were common reasons for admission recorded in the case histories of public patients, and especially those in the armed forces.[14] However, mental symptoms were also recorded in most "service patients": suicide attempts or tendencies, agitation, violence, and irrational ideas.[15] For example, Ioannis Y., while in the military hospital, had exhibited talkativeness ("logorrhoea") and agitation and had been

[13] Vangelis Karamanolakis, "Η ανάδυση των ιδρυματικών θεσμών. Το Δρομοκαΐτειο φρενοκομείο (1880–1903)" ["The emergence of institutions. The Dromokaitio Lunatic Asylum (1880–1903)"], MA thesis, Athens, National and Kapodistrian University of Athens, 1997, pp. 30–3, 62.

[14] *Βιβλίον Ιστορικών [Case Histories Book]*, archive of the Dromokaitio Hospital, Athens (henceforth Dr/Case histories), 1920, pp. 1282–83, 1328; 1918, pp. 956–57; 1919–1920, pp. 1207–8.

[15] Dr/Case histories, 1919–1920, pp. 1207–8; 1920, pp. 1292–93; 1918, pp. 991–92; 1918–1919, pp. 1109–10.

speaking loudly in the ward, saying that he was teaching the lunatics, wanting to "bring them to their senses".[16]

The assessment that "service patients" were suffering from mental illness and would not recover quickly and return to service, as well as their out-of-control symptoms that disrupted the military hospital's routine led to their classification as unsuitable for treatment there. Therefore, the admission of "service patients" to the mental hospitals was probably a decision made by military doctors. In general, it can be said that the hospitalisation of "service patients" in the Eginitio and the Dromokaitio served the public interest in relation to the war effort and the operation of military hospitals. Similar purposes—specialised care and control of people who were causing severe problems in other institutions—can be discerned in the admission of prisoners which was rare. For example, a prisoner who could not be kept in prison because he was too noisy was transferred to the Dromokaitio by order of the Ministry of the Interior.[17]

A more common public function of the hospitals was the admission of civil servants at the request of their superiors. Some of them were considered to have fallen ill because of or in the course of their work: a village constable, admitted to the Dromokaitio by the Ministry of the Interior, had fallen ill after a conflict with draft evaders, and a post office employee, admitted to the Eginitio at the request of his superior, "had fallen ill in service".[18] In addition, civil servants were sometimes admitted when their inability or reluctance to work and the problems they caused at work were attributed to mental illness. For example, the Greek sub-consulate in Kavala, in northern Greece, requested the hospitalisation of an employee because, two months after his appointment, he became agitated and started drinking, which resulted in him performing his duties poorly and causing damages.[19] Dimitrios N., a telegraph operator, was admitted to the Dromokaitio at the request of his superior at the Ministry of the Interior because he rarely went to work, did not explain his absences, only came to ask for his wages in an imperious manner and

[16] Dr/Case histories, 1918–1919, p. 1009.

[17] Anon., "Αι χθεσιναί σκηναί εις το Δρομοκαΐτειον" ["Yesterday's scenes in the Dromokaitio"], *Εμπρός [Forward]*, 8/4/1902, p. 2.

[18] Dr/Case histories, 1918, pp. 908–9; *Clinics-Hospitals*, 1918–1919, 4/4/1919.

[19] Dr/Case histories, 1904–1907, p. 635.

insulted the other employees, the Prime Minister and the King.[20] This case highlights that public admissions were not triggered by an impersonal public sphere, but by specific persons in the patients' environment, such as their superiors or colleagues. It is therefore not always easy to classify admissions as private or public.

The hospitalisation of unmanageable individuals points to another public function of both hospitals: the examination and treatment of people deemed dangerous to themselves or others. For example, 17-year-old Armenia S., who was admitted to the Dromokaitio by the police, was described as having "limited mental capacity" and being unable to look after herself. She refused to eat, had periodic agitation and anxiety, and was incoherent and delusional.[21] Individuals who had been violent and disruptive in public, sometimes under the influence of alcohol, were arrested and brought in by the police, even if they were subsequently admitted at private request.[22] Moreover, the police and other public authorities also intervened in cases where no violence had (yet) occurred. For example, in 1915, a morphine-addicted nurse from a large Athens hospital, the Evangelismos, was sent to the Eginitio on the orders of the public prosecutor and at the expense of the Evangelismos, because she was suffering from "mania of persecution" and kept a revolver in her room.[23] In sum, both psychiatric institutions were occasionally called upon to help maintain public order by admitting patients who were considered offensive and dangerous, but, as already noted, this function of psychiatric hospitalisation, which was provided for by the legislation of the time, was rare.

Finally, the role of public agents in institutional mental health care as supporters of admission applications—especially for free or low-cost hospitalisation—was linked to political motives. A telling example is that of municipal authorities, who were among the main agents in the distribution of free mental health care. This process was fraught with political interests, especially when mayors were trying to secure re-election. For instance, before the municipal elections of 1903 "the offices of the

[20] Dr/Case histories, 1916–1917, pp. 773–74.

[21] Dr/Case histories, 1920–1921, p. 1351.

[22] Eg/9.2, p. 189; Dr/Case histories, 1907–1912, p. 118.

[23] Anon., "Θεραπευτήριον ο 'Ευαγγελισμός'" ["The 'Evangelismos' Sanatorium"], *Εμπρός [Forward]*, 16/2/1915, p. 2.

Dromokaitio Lunatic Asylum were flooded with documents from the Mayor [of Athens] concerning the admission of exceptional patients". Despite the initial objections of the institution, 30 additional municipal patients were admitted, who were gradually discharged after the elections, again on the orders of the municipal authorities.[24] This incident highlights the political uses of public hospitalisation but also confirms that demand for free psychiatric institutionalisation was greater than the supply (see Chapter 4).

In the Eginitio too, public bodies such as the Ministry of the Interior, local authorities, the Church, the university, associations such as the "Macedonian Beneficence" and the "Cretan Brotherhood in Athens", and public figures supported applications for free hospitalisation, in order to promote their charitable and political aims. However, in the case of the university hospital, these efforts were not so often successful, as they came into conflict with the medical director, who had the final say in the selection of pauper patients, based on medical rather than social criteria. As will be shown in Chapter 12, free hospitalisation in the Eginitio triggered heated negotiations between its medical administration and the public, suggesting that the public role of the university hospital was neither self-evident nor as extensive as that of the Dromokaitio. The Eginitio was seen by some state officials as a hospital for the treatment of rich patients with nervous illnesses, and a proposal by a member of parliament a few years after its foundation to turn it into a hospital for the free treatment of paupers suffering from infectious diseases was not accepted.[25] Such a change would have contradicted the scientific and academic rationale behind the establishment of the Eginitio (see Chapter 12).

Public and Private Reactions to Public Admissions

The question explored in the previous chapter, whether those who were institutionalised were truly mentally ill, confined for their care and treatment, was also raised in relation to public patients. Although the public uses of the institutions were viewed as necessary in principle, there was

[24] Anon., "Το σκάνδαλον των οφειλών του δήμου εις το Δρομοκαΐτειον" ["The scandal of the municipality's debts to Dromokaitio"], *Σκριπ [Skrip]*, 13/5/1904, p. 4.

[25] Anon., "Η διπλή Βουλή. Επιψήφισις νομοσχεδίων" ["The double Parliament. Voting on bills"], *Σκριπ [Skrip]*, 18/2/1911, p. 3.

no shortage of accusations of politically motivated "wrongful confinement". In some countries, such as France, from the second half of the nineteenth century there was a strong discourse against asylums, which were presented in the press as prisons used by the state, with the compliance of doctors, to maintain social order and suppress opponents of the regime.[26] This political version of "wrongful confinement" also appeared in public discourse in early twentieth-century Greece. Victims of political persecution and many other "nefarious schemes" were allegedly held in the Dromokaitio[27] and there were occasional calls for the "scientific" examination of its patients to determine whether it was a charitable institution or a "lair of a gang of villains and torturers", a "new Bastille", a "basement of the Holy Inquisition", used to incarcerate people who were sane but inconvenient or dangerous to the state, who had spoken out against the political authorities, and threatened certain political figures.[28] For their part, the doctors dismissed such accusations, attributing them to a misunderstanding on the part of the public and the patients as to the reasons for and purposes of psychiatric confinement.

In the Eginitio and the Dromokaitio, some patients claimed that they had been unfairly and illegally admitted by the public but not for political reasons. Dimitrios N., the post office clerk who was admitted to the Dromokaitio by the Ministry of the Interior at the request of his supervisor, asserted that he was healthy and asked to see his admission certificates, while Nikolaos K., a soldier, stated that he had been admitted to the Dromokaitio in order to be punished by his superiors.[29] In contrast, the private environment of public patients did not seem to react negatively to their admission. Although it is possible that private reactions to public hospitalisation were not recorded, often the private

[26] Jan Goldstein, *Console and Classify: The French Psychiatric Profession in the Nineteenth Century*, Cambridge: Cambridge University Press, 1987, p. 352; Jill Harsin, "Gender, class, and madness in nineteenth-century France", *French Historical Studies*, 17 (1992), pp. 1048–70; Aude Fauvel, "Le crime de Clermont et la remise en cause des asiles en 1880", *Revue d'Histoire Moderne et Contemporaine*, 49 (2002), pp. 195–216; Muriel Bonaïti, "Histoire de l'asile XIXe siècle: D'un lieu unique de soin aux premières alternatives à l'hospitalisation", PhD thesis, Université de Lyon, 2007, pp. 59–61.

[27] Anon., "Τα του φρενοκομείου" ["Issues of the lunatic asylum"], *Εμπρός [Forward]*, 18/10/1901, p. 1.

[28] Anon., "Αι αποκαλύψεις του φρενοκομείου" ["The revelations of the lunatic asylum"], *Εμπρός [Forward]*, 16/10/1901, pp. 1–2.

[29] Dr/Case histories, 1916–1917, pp. 773–74; 1920, pp. 1346–47.

environment of patients was absent from the admission process altogether. Indeed, some public patients had no relatives or friends to care for them and provide information about them.[30] For example, the Prefect of Attica and Boeotia asked that Eleni L. be admitted to the Eginitio free of charge, because the "poor daughter" was "alone and without any protection".[31] For such patients, the very lack of private support, care and supervision networks was one of the factors that motivated their public environment to seek their admission.

In other cases, as already underlined, public hospitalisation was motivated by private initiative and can therefore be assumed that it was regarded positively by at least some members of the patients' private surroundings. A typical example is that of the municipal patients in the Dromokaitio. Initially, a private application was made to the municipality to subsidy the hospitalisation. The applications were discussed in the municipal council and, for those that were approved, the municipality requested admission to the hospital at its own expense.[32] Since the municipal authorities usually found it difficult to pay the fees (see Chapter 4), it is quite unlikely that they decided to send someone to the Dromokaitio unless there was pressure from the patient and his/her environment, which could be translated into political pressure. The case of the municipal patients confirms that the public had more to do with the formal application for admission and the financing of the hospitalisation than with the actual decision and initiation of the hospitalisation.

Public Initiative Discharges

Data on public discharges are scarce because, as already noted, the issues of who decided the end of hospitalisation and why were rarely documented in the hospital archives. In the Dromokaitio, 24% of the 374 patients for whom data are available were discharged by public authorities, usually the Ministries of the Interior, Welfare, the Military or the police. These patients were predominantly public (admitted on public

[30] For instance, Dr/Case histories, 1904–1907, p. 626; 1907–1912, p. 118.

[31] *Clinics-Hospitals,* 1913–1914, 22/8/1914.

[32] The process is described in Anon., "Το χθεσινόν δημοτικόν συμβούλιον" ["Yesterday's city council meeting"], *Σκριπ [Skrip]*, 11/3/1905, p. 1.

request), and some were not discharged but transferred to other institutions, including military hospitals, the public psychiatric hospitals of Corfu and Athens, and prisons. As with admissions, public intervention in discharges did not exclude a role of the patients' private environment. For example, two public patients were discharged to the care of their families at the request of public authorities, while ten of the patients transferred to public psychiatric hospitals at the request of public authorities were private patients. These transfers were usually carried out by the police on the orders of the Ministry of the Interior and rarely at a private request (see Chapter 9); the private environment of the patients was probably not favourable either to the Corfu psychiatric hospital, located on a distant island, or to the public psychiatric hospital of Athens, a police-controlled asylum (see above and Chapter 3).

Some public patients were discharged or transferred for economic reasons. In the Dromokaitio sample, six patients were discharged because their municipal funding was discontinued, which is consistent with information from the press that the municipalities of Athens and Piraeus were trying to reduce the cost of maintaining patients in the Dromokaitio at the beginning of the twentieth century. For example, in 1908, the municipality of Piraeus decided to continue subsidising ten of its fifteen patients, six "incurable" and four "dangerous"; the remaining five, described as "incurable and harmless", would be discharged.[33] A few years earlier, in 1902, the municipality of Athens had carried out an evaluation of the patients it kept in the institution according to similar criteria.[34] Clearly, the local authorities prioritised the hospitalisation of "dangerous" patients, who "for reasons of safety must necessarily be kept there [in the Dromokaitio]", and of curable patients, whose treatment was expected to be shorter and therefore less costly. The rest were to be "sent to their families" or "to Corfu",[35] "since the municipality cannot spend and maintain those with lifelong illnesses".[36] In other words, long-term care was seen as a family responsibility or something to be provided in public hospitals, which were less expensive (see Chapter 3).

[33] Anon., "Η χθεσινή συνεδρίασις του δημ. συμβουλίου Πειραιώς" ["Yesterday's meeting of the Piraeus City Council"], *Εμπρός [Forward]*, 16/3/1908, p. 4.

[34] "Anon., "Επιθεώρησις εις το Δρομοκαΐτειον" ["Inspection at the Dromokaitio"], *Εμπρός [Forward]*, 20/6/1902, p. 2.

[35] Anon., "Yesterday's meeting of the Piraeus City Council".

[36] Anon., "Inspection at the Dromokaitio".

Finally, discharges were carried out on the initiative of the medical authorities of the institutions, usually for patients who were deemed cured or not in need of hospitalisation. For instance, in 1909, the medical director of the Dromokaitio handed Spyridon K. over to his mother after five months in hospital, because he was not "insane". Although he was "certainly a person of unbalanced mind", "mere unbalance" did not justify further hospitalisation, which could be harmful to the patient.[37] In the Eginitio, the medical director discharged patients whom he considered incurable, in his endeavour to create a proper university hospital, namely one for the short-term treatment of curable and acute cases (see Chapter 12). He also discharged patients for whom he recommended other types of treatment: nine patients were transferred from the Eginitio to non-neuropsychiatric institutions, such as a sanatorium and a surgical and gynaecological hospital, without having been treated.

Discharge decisions and criteria were not always accepted by all parties, and controversies arose between the medical authorities of the institutions on the one hand, and the patients and their private environment on the other. The medical authorities could prohibit the discharge of private and public patients if they considered them dangerous or unfit for discharge. In this context, they prevented patients wishing to be discharged from leaving the institution and forbade their relatives and friends to take them out of the institution if they had no legal right to do so. Conversely, the medical directors discharged patients whose relatives wanted them to remain in the hospital. In the Eginitio, pauper patients, especially those deemed incurable, were discharged by the medical director after relatively short intervals sometimes against the wishes of their relatives for treatment and long-term care (see Chapter 12). However, in these disputes the medical authorities did not always prevail over the patients and their private environment. This is illustrated by the most extreme form of confrontation between doctors and patients: escapes (see Chapter 9).

[37] Dr/Register, 1909–1911, year 1909, no 49.

Conclusions

In conclusion, public agents appear to have been less involved in initiating and terminating psychiatric hospitalisation than the patient's private environment. In addition, the public played a lesser role in the hospitalisation decision than in the formal application for admission and the payment of hospital fees. Nevertheless, both institutions, the Dromokaitio to a greater extent than the Eginitio, performed public functions by admitting individuals who publicly showed signs of serious illness or uncontrollable or dangerous behaviour, or who had no or insufficient private networks of support and control. The purpose of public hospitalisation was both to treat and safekeep these patients and to protect society. The public sector, however, was reluctant to fund institutional care—especially long hospital stays—and tended to hospitalise those who were deemed curable or dangerous. It was under these terms that both psychiatric institutions were embedded in wider public networks, serving other institutions, meeting the needs of public authorities for the treatment of patients under their jurisdiction and responsibility, contributing to the war effort, promoting the political activity of public agents and, more rarely, maintaining public order.

When deciding to terminate hospitalisation, public actors took into account their own financial situation and needs, the type of the institution and the patients' interests, at least as each actor understood them. The question whether to discharge or keep a patient, as well as the more general questions of who should be a public patient, who should pay for him/her, and how many such patients there should be, sometimes brought different public agents into conflict with each other or with the patients and their private environment. These conflicts highlight the divergent perceptions of psychiatric hospitalisation and the fluidity between its public and private uses and functions. Indeed, the public and the private were intertwined, as evidenced by private admissions and discharges involving non-private actors, and admissions and discharges that were declared public but were carried out with the decision or cooperation of patients or their relatives. In no case were these actors on an equal footing, but the balance between them was neither self-evident nor fixed, and the actions and aspirations of one affected those of the other.

CHAPTER 11

The Boundaries Between Institutions and Society

The analysis of the public and private uses of institutional mental health care has already shown that psychiatric institutions were not isolated from society: patients and their private and public milieus intervened in different ways and to different degrees in psychiatric hospitalisation, shaping its purposes and determining its beginning and end. This chapter looks in more detail at the relationship between institutions and their social environment by exploring the duration of illness before admission, previous treatments, the length of hospitalisation and, finally, the ways in which institutions were communicating with the outside world.

Duration of Illness Before Admission

Most nineteenth- and twentieth-century psychiatrists advocated early and prompt admission to psychiatric hospitals, arguing that delays in hospitalisation (usually of more than a year) led to failure of treatment and chronicity.[1] This theory was also supported in Greece. Indicatively,

[1] Eric Engstrom, *Clinical Psychiatry in Imperial Germany: A History of Psychiatric Practice*, Ithaca: Cornell University Press, 2003, pp. 32–33; Patricia Prestwich, "Family strategies and medical power: 'Voluntary' committal in a Parisian asylum, 1876–1914", *Journal of Social History*, 1994 (27), pp. 799–814.

D. Kritsotaki, *A Social History of Mental Health Care in Greece, c. 1900–1920*, Mental Health in Historical Perspective,
https://doi.org/10.1007/978-3-032-05764-8_11

Michael Yanniris, the medical director of the Dromokaitio, attributed the low cure rate at the hospital between 1887 and 1902 to the delayed admission of patients, stressing that the family turned to the institution when they had lost all hope of treatment.[2]

Information on the time of onset of illness was usually recorded in the patients' histories, and, on this basis, it can be concluded that most patients were admitted within a year of the manifestation of symptoms, a period considered short by the psychiatrists themselves. Nevertheless, many patients were admitted after longer periods, especially to the Dromokaitio. The difference between the two hospitals could be attributed to the different character of the Eginitio: as a university hospital specialising in neurology, it was less stigmatised and probably discouraged and delayed the admission of patients to a lesser extent. The physiognomy of the institutions therefore had an impact on how they were perceived and used, including how quickly patients were admitted (Table 11.1).

Yet, factors other than the stigma of asylums could delay hospitalisation, such as the tolerance or non-recognition of the first signs of mental illness by the patient's environment. Panagiotis S.'s "first mental disorders"—making business plans and certain "irrational" actions—had been tolerated for more than two years. It was only in the last seven months before his admission that it became clear that he was not well: his condition deteriorated and he had "fully formed ideas of grandeur", such as

Table 11.1 Duration of illness before admission

Onset of illness	*Patients admitted to the Eginitio*		*Patients admitted to the Dromokaitio*	
	N	%	N	%
Up to a year before admission	164	62.4	82	42.1
More than a year before admission	68	25.9	75	38.5
Unknown	31	11.8	38	19.5
Total	263	100.0	195	100.0

Source: Samples from the case histories, Eginitio and Dromokaitio, 1905–1909, 1916–1920 (see Table 1.1).

[2] Dimitris Ploumpidis, *Ιστορία της Ψυχιατρικής στην Ελλάδα. Θεσμοί, Ιδρύματα και Κοινωνικό Πλαίσιο 1850–1920 [History of Psychiatry in Greece: Institutions, Hospitals and Social Framework 1850–1920]*, Athens: Εξάντας [Exandas], 1995, p. 189.

being an emperor and a millionaire.[3] In other cases, the patients' condition was perceived as problematic at an early stage but was tolerated for a long time until it deteriorated. Hospital admission was sought when patients were considered impossible to keep, care for, and supervise at home, especially if their behaviour was deemed dangerous, for example, because of suicidal tendencies.[4] A patient who had been suffering from melancholia for three years and who had been treated at home because his family did not want to seek help from a specialised institution was eventually admitted to the Dromokaitio when he manifested "sitophobia" (an extreme aversion to eating or food) and had not eaten for four days. The "moral means" previously used by his family were no longer working and it was decided that he should be admitted to the hospital for artificial feeding—which usually meant tube feeding.[5] In this account, which does not come from a case history but from a published hospital report, the narratives articulated by the patient's private environment were transcribed into medical language, although it is possible that the patients' relatives were familiar with and used the medical jargon after caring for three years for a diagnosed patient. Lastly, in addition to the deterioration of the patients' condition, changes in their environment, such as the illness or death of their main carer, or the marriage of a close family member, made admission inevitable.[6]

In summary, the timing of admission was determined by the onset and recognition of mental illness symptoms, the degree of patience, tolerance and resilience of the patient's environment, and the attitudes towards the institutions.[7] Within this context, it was not uncommon for hospitalisation to be postponed for a long time because of the reluctance to use

[3] *Βιβλίον Ιστορικών [Case Histories Book]*, archive of the Dromokaitio Hospital (henceforth Dr/Case histories), 1904–1907, p. 712.

[4] Dr/Case histories, 1918, pp. 950–51.

[5] Dromokaitio Hospital, *Απολογισμός και Στατιστική του Δρομοκαϊτείου Φρενοκομείου Έτους 1889 [Report and Statistic of the Dromokaitio Lunatic Asylum for the year 1889]*, Athens: s.n., 1890, pp. 44–45.

[6] *Πρόχειραι Ιστορίαι Νόσου Εισαγομένων Ασθενών Β΄* [*Draft Illness Histories of Patients Admitted B*], archive of the Eginitio Hospital, 9.2, Athens, p. 421 (henceforth Eg/9.2), p. 12.

[7] This has been shown in other countries as well. See, for example, Constance McGovern, "The community, the hospital, and the working-class patient: The multiple uses of asylum in nineteenth century America", *Pennsylvania History*, 54 (1987), pp. 17–33; Prestwich, "Family strategies and medical power".

psychiatric hospitals, especially when these were stigmatised as asylums, and until the patients' condition and the circumstances of those around them had become extremely difficult.

Previous Treatments

The reluctance to use mental institutions is also illustrated by the search for alternative treatments before admission to the Eginitio and the Dromokaitio. Patients had resorted to various non-medical (religious and lay) and medical (empirical and professional) means of healing, including stays in monasteries,[8] rest, holidays and travel,[9] hydrotherapy ("cold showers"), especially for nervous disorders,[10] and remedies from empirical doctors[11] and/or medical treatment in general hospitals or private clinics, especially for physical ailments, such as syphilis.[12] Several had consulted doctors, often as a first step towards more extensive medical intervention. For example, according to the relative of a patient brought to the Dromokaitio from Egypt, once the patient became "quiet, pensive, melancholic" and manifested ideas of persecution the family called the family doctor. Later, when the delusions became more serious and the patient's behaviour more difficult—he was reported to have broken the milk bottle because he thought his relatives were putting poison in it—the family begged this relative who was providing the case history to take the patient to Athens to be admitted to the Dromokaitio. The relative gathered information about doctors in Athens and, when he arrived in the city with the patient, he called the "neurologist" Konstantinos Tsiminakis to the hotel where they were staying. Tsiminakis examined the patient and recommended admission to his private clinic. Three months later, the patient had not improved, and the same relative applied for admission to the Dromokaitio.[13] This case, as well as others like it, illustrates that the pathways to treatment were sometimes complex, but also that

[8] *Πρόχειραι Ιστορίαι Νοσούντων Α΄* [*Draft Histories of The Ill A*], archive of the Eginitio Hospital, 9.1, Athens, pp. 49 and 53 (henceforth Eg/9.1); Eg/9.2, p. 159.

[9] Dr/Case histories, 1918, pp. 986–87.

[10] Eg/9.2, pp. 16, 41.

[11] Eg/9.2, pp. 337, 427.

[12] Eg/9.1, p. 203; Eg/9.2, pp. 365, 377; Dr/Case histories, 1920, pp. 1322–23.

[13] Letter by a relative of the patient Errikos A., Dr/Case histories, 1904–1907.

the medical networks of the Greek capital were known to Greeks outside the Greek state.[14]

In short, many patients arrived at the psychiatric hospitals having exhausted a range of practices for coping with their problems: care at home, in hospitals or private clinics, and a variety of methods of relief and treatment, psychiatric, neurological, medical, lay, and religious. To some extent, these complex therapeutic pathways reflected the reluctance of the family and society towards psychiatric hospitals.[15] More generally, however, they show that the patients moved between different therapeutic sites, which were not mutually exclusive but could be used alternatively, successively or in combination to deal with the difficult and often incomprehensible facets of mental illness.

Duration of Stay and Models of Hospitalisation

If psychiatric hospitals were not always the first option in the search for treatment or care, they were often not the last—after all, many patients were discharged from the Eginitio and the Dromokaitio without being cured (see Table 9.2). The records of the hospitals did not follow the patients after their discharge, but there is evidence that their treatment continued. Some were transferred to other psychiatric hospitals (see Chapters 9 and 10), while others were hospitalised again in the same institution. Of the patients' admitted to the Dromokaitio and the Eginitio, about 12% and 8%, respectively, were not first-time admissions, but were readmitted a few months or even years after being discharged. Some had been admitted several times to the same institution—the most extreme example was that of Evridiki C., who was admitted to the Eginitio ten times.[16] Readmissions were attributed by doctors to the incurable, chronic and periodic nature of many mental illnesses, but they are of interest here because they demonstrate the permeability of the boundaries

[14] For example, letter by I. L., a patient's brother, Dr/Case histories, 1900–1902.

[15] Tomes, *The Art of Asylum-Keeping*, pp. 103–108; Prestwich, "Family strategies and medical power".

[16] Eg/1, no 431.

between the institution and society: institutions could be used repeatedly according to the needs of the patients and their environment, not necessarily once and for all.[17]

Readmissions are linked to another important aspect of institutional treatment that also highlights the porous boundaries of institutions: the incidence of short-stay and recurrent hospitalisation. Contrary to the prevailing image of psychiatric asylums in the nineteenth and early twentieth centuries, before the era of deinstitutionalisation, community psychiatry and psychotropic drugs, most patients did not spend long periods in them, and therefore readmission was not uncommon.[18] In the two institutions studied, there were patients who were hospitalised for many years—the maximum length of stay was 36 years in the Dromokaitio and 23 in the Eginitio—but the majority of patients admitted were discharged within a year of admission (Table 11.2). Recurrent hospitalisation in the same or another institution also occurred, and it offered an alternative to long-term confinement to meet the needs of patients with incurable and intermittent conditions.

Long-term hospitalisation was more common in the Dromokaitio than in the Eginitio, and the average length of stay was about three years in the Dromokaitio and six months in the Eginitio. In the latter, neurological patients were hospitalised for shorter periods than psychiatric patients (116 and 231 days respectively on average), but even psychiatric patients had a much shorter average length of stay than in the Dromokaitio. Thus, the neurological specialisation and the university character of the Eginitio, aimed at treating mainly acute cases, provided a different treatment model, based more on short-term hospitalisation. Therefore, the

[17] See McGovern, "The community, the hospital, and the working-class patient".

[18] Charlotte MacKenzie, "Social factors in the admission, discharge, and continuing stay of patients at Ticehurst Asylum, 1845–1917", in William Frederick Bynum, Roy Porter and Michael Shepherd (eds), *The Anatomy of Madness: Essays in the History of Psychiatry*, v. 2, London: Tavistock, 1985, pp. 147–74; Robert Ellis, "The asylum, the Poor Law, and a reassessment of the four-shilling grant: Admissions to the county asylums of Yorkshire in the nineteenth century", *Social History of Medicine*, 19 (2006), pp. 55–71; Joost Vijselaar, "Out and in: The family and the asylum. Patterns of admission and discharge in three Dutch psychiatric hospitals, 1890–1950", in Marijke Gijswijt-Hofstra, Harry Oosterhuis, Joost Vijselaar and Hugh Freeman (eds), *Psychiatric Cultures Compared: Psychiatry and Mental Health Care in the Twentieth Century: Comparisons and Approaches*, Amsterdam: Amsterdam University Press, 2006, pp. 277–94.

Table 11.2 Length of stay in the Eginitio and the Dromokaitio

Length of stay	*Patients admitted to the Eginitio*		*Patients admitted to the Dromokaitio*	
	N	%	N	%
Up to a year	1328	92.9	577	60.0
Longer than a year	93	6.5%	347	36.1
Unknown	9	0.6%	37	3.9
Total	1430	100.0%	961	100.0
Average	178 days (6 months)		1213 days (3 years)	

Source: Sample from the registers, Eginitio and Dromokaitio, 1905–1909 and 1916–1920 (see Table 1.1).

duration and model of hospitalisation were linked to the type of institution, the condition of the patients and the needs that they and their environment had. The coexistence of the three different models of hospitalisation (short-term, long-term and recurrent) in the Eginitio and the Dromokaitio shows that they were used in different ways: for shorter or longer periods or at intervals.

Relations Between Institutions and the Outside World

Communication between psychiatric institutions and the outside world was certainly not easy or frequent. Their location in the countryside, far from urban centres, was considered a prerequisite for therapeutic success from the end of the eighteenth century, but their isolation cast a veil of mystery over them and facilitated the spread of rumours.[19] Moreover, the isolated location of psychiatric institutions created practical problems. According to the board of directors of the Dromokaitio, the distance of the hospital from the centre of Athens and the lack of public transport made it difficult both for the board of directors to supervise and for relatives to visit, leaving patients feeling like "lepers", "incurable" and abandoned by their loved ones and society. For these reasons, the board of directors requested the extension of the railway to the Dromokaitio

[19] Engstrom, *Clinical Psychiatry in Imperial Germany*, pp. 19, 33.

so that access would be easy and economical.[20] In order to facilitate the relatives' communication with the institution and the admission and discharge of patients, the Dromokaitio maintained an office in the centre of Athens.[21]

Despite the difficulties, which mainly affected the Dromokaitio because the Eginitio was centrally located, the institutions did communicate with the outside world, largely through visits from the officials responsible for supervising them and from the relatives and friends of the patients, as was the case in other countries.[22] There is more evidence for the Dromokaitio, which was visited by prefectural and municipal officials, as required by law 742,[23] but also by journalists in order to convey the image of the institution to the outside world. For example, in 1901, the newspaper *Εμπρός* [*Forward*] sent a correspondent to the Dromokaitio, who walked around the institution, met patients and published his impressions of life there, as well as extracts from his conversations with male and female patients.[24] More generally, the publication of articles about institutions demonstrates society's interest and curiosity about them and

[20] *Λογοδοσία του Διοικητικού Συμβουλίου του εν Δαφνίω Φρενοκομείου Δρομοκαϊτείου, 1898: Πεπραγμένα [Report of the Board of Directors of the Dromokaitio Lunatic Asylum in Daphni, 1898: Activities]*, Athens: s.n., 1899, pp. 12–13.

[21] Dromokaitio informative leaflet, 1914, reproduced in Nikolaos Tsikis (ed.), *Δρομοκαΐτειο Θεραπευτήριο, 1887–1987 + 15. Εκατό Δεκαπέντε Χρόνια Κοινωνικής Προσφοράς [Dromokaitio Sanatorium, 1887–1987 + 15. One Hundred Fifteen Years of Social Offer]*, Athens: s.n., 2001, p. 59.

[22] Graham Mooney and Jonathan Reinarz (eds), *Permeable Walls: Historical Perspectives on Hospital and Asylum Visiting*, Amsterdam: Rodopi, 2009; Janet Miron, *Prisons, Asylums, and the Public: Institutional Visiting in the Nineteenth Century*, Toronto: The University of Toronto Press, 2011.

[23] For instance, Anon., "Η επίσκεψις του Νομάρχου εις το Δρομοκαΐτειον" ["The Prefect's visit to Dromokaitio"], *Εμπρός [Forward]*, 18/10/1901, p. 2; Anon., "Ο κ. Νομάρχης εις το Δρομοκαΐτειον και το Εφηβείον" ["The Prefect at the Dromokaitio and the Adolescents' Reformatory"], *Εμπρός [Forward]*, 5/2/1902, p. 3; Anon., "Επιθεώρησις εις το Δρομοκαΐτειον" ["Inspection at the Dromokaitio"], *Εμπρός [Forward]*, 20/6/1902, p. 2. However, it has been disputed whether Greek mental hospitals were subject to the required systematic official supervision (Vangelis Karamanolakis, "Η ανάδυση των ιδρυματικών θεσμών. Το Δρομοκαΐτειο φρενοκομείο (1880–1903)" ["The emergence of institutions. The Dromokaitio Lunatic Asylum (1880–1903)"], MA thesis, Athens, National and Kapodistrian University of Athens, 1997, p. 61).

[24] Anon., "Αι αποκαλύψεις του φρενοκομείου" ["The revelations of the lunatic asylum"], *Εμπρός [Forward]*, 18/10/1901, pp. 1–2.

their communication, however limited and fragmentary, with the outside world.

The most common form of contact between institutions and society was visits and correspondence between the patients and their relatives and friends. These means of communication were not always regular, possible or unrestricted. Many letters did not reach their destination—some are contained in case books to this day—while the contemporary press occasionally reported incidents of obstructed visits from relatives. For example, in 1902, there was trouble at the Dromokaitio when a patient's nephew was not allowed to see him; the patient became enraged and beat up the guard and the assistant director of the institution.[25] Thus, although the authorities of the institutions were able to interfere with communication with the outside world, they did not have total control over it, nor did they go unchallenged.

Conclusions

In conclusion, the delay in hospitalisation, the attempt to find other ways of coping with mental illness before admission, and the high extent of involuntary hospitalisation (see Chapter 9) indicate that institutions could appear impenetrable to patients and their environment. This led to reluctance to approach and use them, particularly when combined with the limited number of places, the high cost of hospitalisation, and the difficulty in recognising and accepting patients' problems as symptoms of mental illness. However, this picture is tempered by the existence of opportunities for communication between institutions and society, as well as the generally short length of hospital stays, the relatively high rates of rapid admission, the readmissions, and some of the characteristics of hospitalisation analysed in the previous two chapters: the private and public control of admissions and discharges, even against medical advice; probationary discharges at the request of patients or their relatives; and escapes. Furthermore, the appropriation of the institutions by

[25] Initially, it was claimed that the trouble was caused when the patient's relatives tried to release him (Anon., "Αι χθεσιναί σκηναί εις το Δρομοκαΐτειον" ["Yesterday's events in the Dromokaitio"], *Εμπρός [Forward]*, 8/4/1902, p. 2), but these assumptions were refuted the next day (Anon., "Αι σκηναί του Δρομοκαϊτείου" ["The events in the Dromokaitio"], *Εμπρός [Forward]*, 9/4/1902, p. 2).

their environment was highlighted by the fact that the type of hospitalisation (short-term, long-term or recurrent) depended not only on the physiognomy of each institution, but also on the condition of the patients and the needs and purposes of their milieu. All in all, the institutions were not completely cut off from society: their boundaries were to some extent permeable, while their uses corresponded to and were adapted to the perceptions and realities of their social environment.

CHAPTER 12

Scientific and Professional Aspects of Institutional Mental Health Care

The previous chapters examined the medical profession among the public agents involved in psychiatric hospitalisation: institutional doctors oversaw the processes of admission and discharge and implemented diagnosis and treatment. Although patients, their private and public environment and the institutions' administration limited medical authority by making decisions about the initiation and termination of hospitalisation, the medical staff, led by the medical director, tended to be in a stronger position. At the same time, the institutions were not only places for the care, treatment and control of patients, but also for the fulfilment of the scientific and professional purposes of psychiatry, mainly research and teaching. This chapter focuses on these aspects of psychiatric hospitalisation.

Institutions and the Organisation of Psychiatry

From the end of the eighteenth century, as medicine moved towards the clinical-anatomical method, which was based on experience and observation, hospitals became the epicentre of medical theory and practice.[1] At the same time, the mental asylum played a similar, if not more important

[1] Michel Foucault, *The Birth of the Clinic: An Archaeology of Medical Perception*, transl. A. M. Sheridan, London: Tavistock, 1976, pp. 17, 83. Certainly, the expansion of hospitals from the eighteenth century onwards was linked not only to scientific changes but also

D. Kritsotaki, *A Social History of Mental Health Care in Greece, c. 1900–1920*, Mental Health in Historical Perspective,
https://doi.org/10.1007/978-3-032-05764-8_12

role for psychiatry: it emerged as the main professional space for psychiatrists and more or less became the cradle of the psychiatric discipline as we know it today.[2] The link between asylums and the professionalisation of psychiatry is illustrated by the fact that the first professional organisations of psychiatrists in different countries were organisations of asylum doctors, such as the British Association of Medical Officers of Asylums and Hospitals for the Insane, founded in 1841, and the Association of Medical Superintendents of American Institutions for the Insane, founded in Philadelphia, in 1844.[3]

In addition to promoting the professional organisation of psychiatry, asylums carried out statistical and laboratory research into the causes and treatments of mental illness and the medical and social characteristics of patients and, as early as the first half of the nineteenth century, were used for the clinical training of medical students and doctors, thanks to their large number of patients.[4] Although some doctors believed that clinical teaching hindered the recovery of patients, from the late 1850s there was increasing support that students should be trained in psychiatric institutions and that the presence of students would also benefit

to the social conditions of the time and the organisation of the medical profession. Lindsay Granshaw and Roy Porter (eds), *The Hospital in History*, London: Routledge, 1989; Lindsay Granshaw, "The rise of the modern hospital in Britain", in Andrew Wear (ed.), *Medicine in Society: Historical Essays*, Cambridge: Cambridge University Press, 1992, pp. 197–218.

[2] Edward Shorter, *A History of Psychiatry, From the Era of the Asylum to the Era of Prozac*, New York: John Wiley and Sons, 1997, pp. 8–17.

[3] Edward Renvoize, "The Association of Medical Officers of Asylums and Hospitals for the Insane, the Medico-psychological Association, and their presidents", in German Berrios and Hugh Freeman (eds), *150 Years of British Psychiatry, 1841–1991*, London: Gaskell, 1991, pp. 29–78; Nancy Tomes, "A generous confidence: Thomas Story Kirkbride's philosophy of asylum construction and management", in Andrew Scull (ed.), *Madhouses, Mad-doctors, and Madmen: The Social History of Psychiatry in the Victorian Era*, London: Athlone Press, 1981, pp. 120–43.

[4] Eric Engstrom, *Clinical Psychiatry in Imperial Germany: A History of Psychiatric Practice*, Ithaca: Cornell University Press, 2003, pp. 30 and 47; Mark Micale, "The Salpêtrière in the age of Charcot: An institutional perspective on medical history in the late nineteenth century", *Journal of Contemporary History*, 20 (1985), pp. 703–31; Shorter, *A History of Psychiatry*, p. 81; Engstrom, *Clinical Psychiatry in Imperial Germany*, p. 47; Renvoize, "The Association of Medical Officers"; John Todd and Lawrence Ashworth, "The West Riding Asylum and James Crichton-Browne, 1818–1876", in German Berrios and Hugh Freeman (eds), *150 Years of British Psychiatry, 1841–1991*, London: Gaskell, 1991, pp. 389–418.

patients.[5] By this time many asylum doctors believed that asylums should be educational centres linked to universities and medical schools, which were called upon to organise and make compulsory the theoretical and clinical teaching of psychiatry.[6] In this way, psychiatrists aimed not only to improve psychiatric training and practice but also to raise the status of psychiatry among medical specialties.[7]

Indeed, when psychiatry was first introduced into university studies at the beginning of the nineteenth century, classes were not systematic, compulsory or extensive and were mainly theoretical. Clinical teaching took place largely in asylums, at least until the middle of the century, when university psychiatry began to be organised.[8] Wilhelm Griesinger (1817–1868), who became professor of psychiatry in Berlin in 1865, argued that asylums were unsuitable for teaching and research and set up a university psychiatric clinic at the Charité hospital in Berlin, inaugurating the tradition of psychiatric clinics that spread throughout Europe. These clinics, unlike asylums, were not primarily concerned with care and treatment, but with laboratory and clinical research and the training of students based on their patients, who were admitted without many formal procedures and hospitalised for short periods to ensure a variety of cases, as the clinics were much smaller than asylums.[9] In general, the clinics were more like hospitals than asylums: they were smaller, centrally located in cities, and combined psychiatry and neurology on the basis that mental illness was a

[5] Engstrom, *Clinical Psychiatry in Imperial Germany*, pp. 44–50; Renvoize, "The Association of Medical Officers". The opening of asylums to students had already been called for in 1830 by John Conolly in *An Inquiry Concerning the Indications of Insanity* (Richard Hunter and Ida Macalpine, *Three Hundred Years of Psychiatry, 1535–1860: A History Presented in Selected English Texts*, London: Oxford University Press, 1963, pp. 806–7).

[6] Engstrom, *Clinical Psychiatry in Imperial Germany*, p. 49.

[7] Renvoize, "The Association of Medical Officers".

[8] Shorter, *A History of Psychiatry*, pp. 69–73; Albrecht Hirschmüller and Magda Whitrow, "The development of psychiatry and neurology in the nineteenth century", *History of Psychiatry*, 10 (1999), pp. 395–423. In France, the first university chair in psychiatry (Clinique de Pathologie Mentale et des Maladies de l'Encéphale) was created in 1877 at the Sainte Anne hospital. Pierre Deniker and Jean-Pierre Olie, *Fou, Moi? La Psychiatrie Hier et Aujourd'hui*, Paris: O. Jacob, 1998, p. 56.

[9] Engstrom, *Clinical Psychiatry in Imperial Germany*, pp 12, 57, 61, 72, 104; Shorter, *A History of Psychiatry*, pp. 73–76.

disease of the brain.[10] In addition, they were promoted as leading scientific institutions and had higher cure rates than asylums, mainly because they gave priority to early treatment of acute and mild cases, which had a better recovery rate than the chronic and severely ill patients usually treated in asylums.[11]

Consequently, university clinics ruptured the geographical and scientific isolation of psychiatry as practised in asylums and were an important means of strengthening psychiatry as a scientific and academic discipline. At the same time, the clinics created a different, more elevated public image of psychiatry, often in combination with neurology, and thus contributed to the expansion of the psychiatric field to include patients who had previously been examined and treated not in asylums but in private clinics, spas, and the private practices of neurologists and pathologists. These were patients with mild mental illness or nervous disorders and/or who felt that asylums were not compatible with their social and economic status.[12]

At the beginning of the twentieth century, the two models of mental health care and professional organisation of psychiatry, the asylum and the university clinic, coexisted, and sometimes controversy arose between asylum and clinic doctors over the effectiveness and reliability of each.[13] Both, however, served the professional objectives of psychiatry, albeit in different ways and to different degrees, as will be shown below with the help of the Dromokaitio, which was closer to the asylum model, and the Eginitio, a university clinic.

Research and Teaching at the Dromokaitio and the Eginitio

Nineteenth-century Greek mental hospitals, namely the Corfu Psychiatric Hospital and the Dromokaitio, emphasised that they served the scientific treatment of mental illness (see Chapter 3), but it is not known to

[10] William Bynum, "The nervous patient in 18th- and 19th-century Britain: The psychiatric origins of British neurology", in Robin Murray and Trevor Turner (eds), *Lectures on the History of Psychiatry: The Squibb Series*, London: Gaskell, 1990, pp. 115–27.

[11] Engstrom, *Clinical Psychiatry in Imperial Germany*, pp. 2, 57, 90–91.

[12] Shorter, *A History of Psychiatry*, ch. 4.

[13] Engstrom, *Clinical Psychiatry in Imperial Germany*, pp. 2, 26, 51, 60, 73, 104.

what extent they carried out research and teaching. In the late nineteenth century and until the establishment of the Eginitio, medical students visited the Dromokaitio to observe patients,[14] and its medical director published medical observations in the hospital's annual reports, with the aim, as he noted, of stimulating the cultivation of "the phreniatric clinic".[15] In 1898, the then assistant director and later director of the Dromokaitio, Michael Yanniris (1865–1956), published a study based on his statistical and clinical observations,[16] and in the following years he continued to research and publish mainly on general paralysis.[17] It can thus be argued that the Dromokaitio was active in teaching and research.[18] However, from 1904, the newly established university

[14] Vangelis Karamanolakis, "Η ανάδυση των ιδρυματικών θεσμών. Το Δρομοκαΐτειο φρενοκομείο (1880–1903)" ["The emergence of institutions. The Dromokaitio Lunatic Asylum (1880–1903)"], MA thesis, Athens, National and Kapodistrian University of Athens, 1997, p. 56.

[15] Dromokaitio Hospital, *Απολογισμός και Στατιστική του Δρομοκαϊτείου Φρενοκομείου Έτους 1889 [Report and Statistic of the Dromokaitio Lunatic Asylum for the year 1889]*, Athens: s.n., 1890, p. 35.

[16] Michael Yanniris, *Αι Φρενοπάθειαι εν Ελλάδι: Μελέτη επί Στατιστικών και Κλινικών Παρατηρήσεων [The Phrenopathies in Greece: A Study on Statistical and Clinical Observations]*, Syros: s.n., 1898.

[17] Michael Yanniris, *Περί Προϊούσης Καθολικής Παραλύσεως: Μελέτη Κλινική και Ιατροδικαστική [On General Paralysis of the Insane: Clinical and Forensic Study]*, Athens: s.n, 1900; Michael Yanniris and Nikolaos Arkalidis, "Η δι' ελονοσίας θεραπεία της προϊούσης καθολικής παραλύσεως" ["Malarial treatment of general paralysis of the insane"], in *Πανηγυρικόν Τεύχος επί τη Τριακονταπενταετηρίδι Μιχαήλ Κατσαρά: Εκδοθέν υπό Συναδέλφων και Μαθητών αυτού, 29 Μαρτίου 1928 [Commemorative Issue on the Thirty-fifth Anniversary of Michael Katsaras: Published by his Colleagues and Students, 29 March 1928]*, Athens: Π. Δ. Σακελλάριος [P.D. Sakellarios Publishing House], 1928, pp. 161–79; Michael Yanniris, "Η σύφιλις και αι ψυχοπάθειαι και η διαγνωστική αξία των βιολογικών ερευνών" ["Syphilis and psychopathies and the diagnostic value of biological investigations"], in *Πανηγυρικόν Τεύχος επί τη Τριακονταπενταετηρίδι Μιχαήλ Κατσαρά: Εκδοθέν υπό Συναδέλφων και Μαθητών αυτού, 29 Μαρτίου 1928 [Commemorative Issue on the Thirty-fifth Anniversary of Michael Katsaras: Published by his Colleagues and Students, 29 March 1928]*, Athens: Π. Δ. Σακελλάριος [P.D. Sakellarios Publishing House], 1928, pp. 83–87.

[18] Michalis Yanniris, "Η επιστημονική δραστηριότητα του Δρομοκαΐτειου στα πρώτα πενήντα χρόνια της λειτουργίας του" ["The scientific activity of the Dromokaiteio in the first fifty years of its operation"], in Nikolaos Tsikis (ed.), *Δρομοκαΐτειο Θεραπευτήριο, 1887–1987 + 15. Εκατό Δεκαπέντε Χρόνια Κοινωνικής Προσφοράς [Dromokaitio Sanatorium, 1887–1987 + 15. One Hundred Fifteen Years of Social Offer]*, Athens: s.n., 2001, pp. 95–103.

hospital, the Eginitio, began to play the central professional and scientific role in Greek psychiatry.

The teaching of "phrenic and nervous diseases" had begun at the University of Athens in 1852, but the first professors were not specialised in psychiatry,[19] as was the case in other countries at that time.[20] The neuropsychiatric specialty began to emerge in Greece in the last decade of the nineteenth century: in 1891 the position of professor of "neurology and phrenic diseases" was created at the University of Athens (the only Greek university at the time) and Michael Katsaras (1860–1939) was appointed to it in 1893. Katsaras remained professor of neurology and psychiatry and director of the university clinic until 1931 (see Chapter 3).[21] He also wrote the first Greek textbook on neurology and psychiatry.[22] In short, at the University of Athens Medical School, psychiatry had been taught in combination with neurology, on a regular basis and by specialised staff from the end of the nineteenth century.

Within the context of the increasing specialisation and organisation of psychiatry and neurology, clinical teaching at the City Clinic (Astykliniki), the Athens Municipal Hospital and the Dromokaitio was considered insufficient in the late nineteenth century. At that time, the university seized the opportunity to establish a university hospital with the bequest of professor Dionysios Eginitis, and Katsaras proposed that the new hospital should house a neurological clinic, arguing that "the subject of neurology" was "important" and "eminently practical" and that its theoretical teaching alone was of no use.[23] In addition to clinical teaching, which would be based on paupers, i.e. those who were hospitalised free

[19] Neurological Clinic, University of Athens, *Αιγινήτειον Νοσοκομείον, 1905–2005 [Eginitio Hospital, 1905–2005]*, Athens: s.n., 2005, p. 9; University of Athens, *Επετηρίδες των Πανεπιστημιακών Ετών 1864–1865–1881–1890 [Yearbooks of the Academic Years 1864–1865–1881–1890]*, Athens, 1865–1891.

[20] Shorter, *A History of Psychiatry*, pp. 71–73.

[21] Katsaras was first appointed to the university in 1884–1885 as a lecturer in special pathology, teaching the courses of nervous diseases, spinal cord diseases and special pathology (mental diseases). University of Athens, *Yearbooks of the Academic Years 1884–1885–1892–1893*, Athens, 1885–1894.

[22] Michael Katsaras, *Παθολογία των Νεύρων και Ψυχιατρική [Pathology of the Nerves and Psychiatry]*, 3 vols, Athens: Εστία [Estia], 1898–1900.

[23] "Πρακτικά της συνεδρίασης της πρυτανείας, 11/12/1899" ["Minutes of the meeting of the rectorate, 11/12/1899"], Neurological Clinic, University of Athens, *Eginitio Hospital*, pp. 12–13.

of charge or paid reduced third-class hospital fees, the new hospital, Eginitio, would also provide paid hospitalisation,[24] which Katsaras saw mainly as a source of income for the institution.[25] In other words, at least for the director of the hospital, the Eginitio was primarily intended to respond to the needs of academic psychiatry through the hospitalisation of paupers, while paid hospitalisation was primarily a means of maintaining the hospital. The emphasis on clinical teaching was also laid down in law 3823 "On the Organisation of the National University", which stipulated that at least thirty free beds and ten third-class beds should be available for teaching in each clinic of the Medical School.[26] For the Eginitio, this meant that 40 of the 70 beds would be used for teaching, while the other patients would not exceed 30.[27]

It soon became clear, however, that this ratio of pauper to private patients could not be maintained for financial reasons. Although, according to its regulation, the Eginitio was to receive an annual subsidy from the university, and the hospital's board of administration had applied to the senate for an annual subsidy of at least 8,000 drachmas to maintain the free patients, the institution had to be financially self-sufficient.[28] This meant that it had to rely on payments from private patients to cover its

[24] "Περί εγκρίσεως Οργανισμού του Αιγινήτειου Νοσοκομείου" ["On the approval of the Statute of the Eginitio Hospital"], *Εφημερίς της Κυβερνήσεως [Government Gazette]*, 169, 30/7/1904.

[25] "Minutes of the meeting of the rectorate, 11/12/1899". From the middle of the nineteenth century clinical teaching was typically carried on pauper patients and all the hospitals where medical students were taught, i.e. the Municipal Hospital of Athens, the City Clinic and the Eye Clinic, were public institutions that mainly treated the indigents. Maria Korasidou, *Όταν η Αρρώστια Απειλεί: Επιτήρηση και Έλεγχος της Υγείας του Πληθυσμού στην Ελλάδα του 19ου Αιώνα [When the Disease Threatens: Surveillance and Control of Population Health in 19th-Century Greece]*, Athens: Τυπωθήτω [Tipothito], 2002, pp. 75–76, 81–84). In the Dromokaitio, too, students could only visit the pauper patients (Karamanolakis, "The emergence of institutions", p. 56).

[26] Νόμος ΓΩΚΓ' "Περί Οργανισμού του Εθνικού Πανεπιστημίου" [Law 3823 "On the Statute of the National University"], *Εφημερίδα της Κυβερνήσεως [Government Gazette]*, 178, 12/7/1911.

[27] Letter dated 12/9/1912, *Κλινικές-Νοσοκομεία [Clinics-Hospitals]*, 1911–1912, Archive of the Rectorate of the University of Athens, Historical Archive of the University of Athens, Athens.

[28] Ibid.; Neurological Clinic, University of Athens, *Eginitio Hospital*, p. 13; Anon., "Αι πανεπιστημιακαί κλινικαί και οι φοιτηταί της Ιατρικής" ["University clinics and medical students"], *Σκριπ [Skrip]*, 5/12/1909, p. 2.

expenses, including the maintenance of pauper patients. It is telling that, in order to have enough private patients to support the pauper patients, Katsaras treated some of his own patients not in his private clinic but in the first and second classes of the Eginitio.[29] However, the budgetary needs of the institution and the financial problems that had already arisen in the 1910s did not allow for the admission of the originally planned number of paupers. A year after the publication of law 3823, when the number of beds in the Eginitio were almost 100, Katsaras informed the rector that each paying class had about 35 patients to cover the hospital's expenses, while there were only ten pauper patients, the same number as since the foundation of the Eginitio, when it treated a total of only 24 patients.[30] In quantitative terms at least, the educational purpose of the hospital was undermined, and the medical director, trying to cope with the hospital's persistent financial difficulties, rejected requests for reduced fees,[31] which did not help the institution financially, on the contrary, created deficits.[32]

Since free beds were steadily limited, their allotment was a contested issue. Free or reduced-fee hospitalisation was claimed by patients, their private and public environment, and the medical authority of the hospital. The family, as well as ministries, prefectures, municipalities, communities, the Church, various organisations, and public figures, asked the Eginitio to play a social role by helping patients in dire economic and social circumstances (see Chapter 4), including refugees, patients without family, patients who had served their country or suffered in wars, and university students or graduates who believed they were entitled to preferential treatment.[33] In general, the applicants appealed to the sensibilities of the Eginitio authorities, asking for the rescue and relief of unfortunate

[29] "University clinics and medical students".

[30] Letter dated 12/9/1912, *Clinics-Hospitals*, 1911–1912.

[31] For example, see letters dated 19/12/1914, *Clinics-Hospitals*, 1914–1915, and 27/7/1916, *Clinics-Hospitals*, 1915–1916.

[32] For instance, in 1913–1914 the hospitalisation of patients paying reduced fees had caused a deficit of 2,000 drachmas. Letter dated 19/12/1914, *Clinics-Hospitals*, 1914–1915.

[33] Letters dated 27/11/1908, *Clinics-Hospitals 1908–1909*; 9/12/1910 and 18/1/1911, *Clinics-Hospitals*, 1910–1911; 14/7/1912, *Clinics-Hospitals*, 1911–1912; 22/8/1914, *Clinics-Hospitals*, 1913–1914; 20/11/1914, 26/2/1915 and 28/4/1915, *Clinics-Hospitals*, 1914–1915; 20/9/1916, *Clinics-Hospitals*, 1916–1917; 13/4/1920, *Clinics-Hospitals*, 1919–1920.

and poor patients and families, not only through short-term treatment but also though long-term care.[34] The rector, who received the applications (see Chapter 10), supported some of them on the basis of social criteria. For example, when he recommended Maria P. for free treatment, he emphasised the "valuable services" she had rendered to the "struggles" of Crete for the unification with Greece.[35] The rector also took into account who was making the request or acted as a guarantor for the patient. The application for free hospitalisation of the reserve sergeant Alexander D. was accepted after the rector insisted on pleasing the patient's "patroness", the Queen,[36] while the request from the Athenian monastery of Asomaton-Petraki for a reduction in the fees of one of its monks was accepted after the monastery stressed its contributions to the university and its clinics.[37]

The efforts of patients and their private and public environment to use the Eginitio as a charitable institution offering free treatment according to social criteria met with the opposition of its medical director, who claimed free hospitalisation for teaching and research. He sought to have complete control over the selection of paupers, the "teaching material", as he called them.[38] Indeed, the admission of pauper patients required his permission, which was not necessary for the admission of fee-paying patients. As the director's letters to the rector show, the criteria used by the director to make his selection were educational and scientific. To be admitted free of charge, a patient had to have either a curable disease or at least one with the potential for improvement or a condition of educational interest.[39] The rejection of patients with chronic and incurable diseases was particularly common. The director believed that difficult incurable cases belonged in the asylum for the incurable,[40] while he argued that quiet and harmless chronic patients could and should stay at home, as

[34] Letter dated 7/10/1915, *Clinics-Hospitals,* 1915–1916.

[35] Letter dated 20/10/1917, *Clinics-Hospitals,* 1917–1918.

[36] Letters dated 19/2/1917 and 29/4/1917, *Clinics-Hospitals,* 1916–1917.

[37] Letter dated 27/4/1909, *Clinics-Hospitals,* 1908–1909.

[38] Letter dated 12/9/1912, *Clinics-Hospitals,* 1912–1913.

[39] Letters dated 27/5/1911, *Clinics-Hospitals,* 1910–1911; 2/10/1917 and 14/10/1917, *Clinics-Hospitals,* 1917–1918.

[40] Letter dated 27/5/1917, *Clinics-Hospitals,* 1917–1918. For the Asylum for the Incurable see Chapter 3.

they would be harmed by hospital isolation.[41] For example, in a letter to the rector, Katsaras explained that he had sent home a patient suffering from "congenital stupidity", because he was quiet and would benefit more from "family care" than from further hospitalisation, which would deteriorate his character and "remaining small mental capacity" due to the negative influence of patients with serious illnesses.[42] Once again the public environment of patients—this time the medical authority of the Eginitio—presented family care as the main care option for chronic but manageable patients (see Chapter 10).

The criteria for free hospitalisation are summarised in a letter from Nikolaos Makkas, director of the Eginitio in 1911, which reveals the diseases that interested the hospital's medical authority. Makkas requested the discharge of a female patient, a "lunatic" who had been hospitalised free of charge for two months, arguing that "psychopaths" were not admitted for free and that this patient had shown no improvement.[43] In short, the medical director favoured the selection of paupers who were suffering from neurological conditions or who at least had some prospect of improvement.[44] Evidently, the hospital's medical authority sought to control who was hospitalised for free, in order to shape and serve the scientific profile and objectives of the Eginitio.

[41] For example, letters dated 1/8/1914, *Clinics-Hospitals,* 1913–1914; 3/6/1917, *Clinics-Hospitals,* 1916–1917.

[42] Letter dated 20/12/1912, *Clinics-Hospitals,* 1912–1913.

[43] Letter dated 27/5/1911, *Clinics-Hospitals,* 1910–1911. Makkas replaced Katsaras for a year (see Chapter 3), but he shared the same criteria for selecting paupers with the hospital's long-serving director.

[44] The interest in neurology was linked to the scientific interests of Katsaras, the professor who laid the scientific foundations of the hospital and was its director for almost the entire period under consideration. After graduating from the Athens Medical School, Katsaras trained in Paris not only in psychiatry but also in neurology, notably at the Salpêtrière hospital from 1880 to 1883 under Jean-Martin Charcot who was known for his studies and courses on nervous diseases such as hysteria, multiple sclerosis and chorea (Michael Katsaras, *Παθολογία των Νεύρων και Ψυχιατρική. Νευρώσεις, Νευρίτιδες, Περιφερικαί Παραλύσεις [Pathology of the Nerves and Psychiatry. Neuroses, Diseases of the Nerves, Peripheral Paralyses]*, v. 3, Athens: Εστία [Estia], 1900, p. 8). Katsaras' doctoral thesis, submitted in 1883, was on multiple sclerosis, while many of his subsequent studies were "nervous diseases" (Aristotle Kouzis, *Εθνικό και Καποδιστριακό Πανεπιστήμιο Αθηνών, Εκατοενταετηρίς 1837–1937*, *Γ′ Ιστορία της Ιατρικής Σχολής [National and Kapodistrian University of Athens, Centenary 1837–1937, v. 3 History of the Medical School]*, Athens: Τύποις "Πυρσού" [Pyrsos Publishing House], 1939, pp. 32–33).

Within this context, when the medical director gave way to applications with significant guarantors, he usually imposed conditions. For example, the application of Alexander D. mentioned above was accepted thanks to his guarantor, but on the condition that the patient would be admitted a month later, when there would be a vacant pauper's place, and that he would be discharged if he was found to be incurable.[45] This case illustrates that compromises were taking place between the rector's and the medical director's approach to free hospitalisation as a social and scientific instrument, respectively.[46] On the contrary, at other times the two sides came into conflict. For example, in 1915 Katsaras emphatically reminded the senate that the ten beds for the paupers had an "exclusively educational purpose" and not a "charitable" one, and that it was up to the professor to decide which cases were useful for teaching. He therefore asked the senate not to use the ten beds for charitable purposes, because this would interfere with teaching.[47]

The scientific claims of the Eginitio were known and emphasised by its academic and social environment. In his annual speech for 1904–1905, the year of the hospital's opening, the rector focused on the use of scientific methods—specifically hydrotherapy and electrotherapy—for "treating diseases that are exceptionally prevalent in our country due to the extraordinarily nervous character of our nation".[48] Moreover, some applicants for free hospitalisation appealed to the scientific quality of the institution, calling for "systematic scientific treatment" in a "special neurological clinic" or referring to Katsaras' "wonderful scientific ability".[49] On the other hand, the medical authority of the hospital recognised society's needs for treatment: it presented the Eginitio as an institution of "maximum social benefit", a "unique example of a large clinic for nervous and

[45] Letter dated 3/5/1917, *Clinics-Hospitals,* 1916–1917.

[46] The medical director also made compromises with public agents requesting free hospitalisation. For example, in 1916, Katsaras accepted the admission of a patient by the police on the condition that the patient would be discharged if he proved to be incurable. Letter dated 22/7/1916, *Clinics-Hospitals,* 1915–1916.

[47] Letter dated 19/10/1915, *Clinics-Hospitals,* 1915–1916.

[48] National University, *Τα κατά την Πρυτανείαν Σπυρ. Π. Λάμπρου Πρυτανεύσαντος κατά το Ακαδημαϊκόν Έτος 1904–1905 [The Activities of the Rectorate of Spyr. P. Lambros Rector during the Academic Year 1904–1905]*, Athens: Τύποις Π.Δ. Σακελλαρίου [P.D. Sakellarios Publishing House], 1907, pp. 14–15.

[49] Letters dated 24/7/1914, *Clinics-Hospitals,* 1913–1914; 20/9/1916, *Clinics-Hospitals,* 1916–1917; 11/1/1916, *Clinics-Hospitals,* 1915–1916.

mental diseases", necessary for the treatment of patients who could not be treated at home or in other institutions, since the Dromokaitio had a "huge number" of patients and the private clinics in Athens were full.[50]

Thus, each side was to some extent aware of the needs of the others, but still focused on its own needs. On the one hand, the rector did not want to displease the public authorities and influential guarantors by excluding the patients they supported, while most of the private and public applicants did not consider or were unaware of the teaching purposes and medical criteria of free hospitalisation. On the other hand, the medical director appeared to overlook or not understand the problems that led patients and their environment to seek institutional care, in particular free and long-term hospitalisation (see Chapters 9 and 10). Indicatively, the director could not understand why the family refused to care for quiet and incurable patients who, in his opinion, would benefit more from family than institutional care,[51] and declared that it was "impossible for the Eginitio with its ten beds of paupers to solve the problems of the 6,000 mentally ill of the Greek state".[52]

Who were the ultimate winners in the struggle for free hospitalisation in the university hospital? The profile of the paupers, the category that included both free and third-class reduced-fee patients, can help to answer this question. According to the patient registers, the paupers of the Eginitio, like the paying (or "well-off" as they were called) patients, were mainly male, adult and unmarried, and usually belonged to the middle and lower middle social classes, i.e. they were craftsmen and shopkeepers with small businesses, employees, and farmers.[53] On the other hand, there are some differences in the social characteristics of the two groups. The paupers had a higher proportion of labourers and unskilled workers, such as cleaners and ushers, while the well-off had a higher proportion of senior civil servants, landowners, and professionals, such as engineers, lawyers, and doctors. In addition, the paupers were more likely to be under 21, unmarried, and widowed than the well-off (see Tables 12.1, 12.2 and 12.3). Thus, the paupers belonged to more vulnerable economic and social groups and were more likely to lack family support and to be

[50] Letter dated 20/4/1911, *Clinics-Hospitals,* 1910–1911.

[51] Letter dated 20/12/1912, *Clinics-Hospitals,* 1912–1913.

[52] Letter dated 11/1/1916, *Clinics-Hospitals,* 1915–1916.

[53] On professions and social class, see Chapter 7.

unable to pay for hospital care. However, the analysis of their social characteristics supports the hypothesis that the term "pauper" in psychiatric institutions should not be understood as "poor".[54] Indeed, although the pauper patients of the Eginitio were supposed to be "destitute",[55] they were a category with internal differentiations and quite similar in social terms to the well-off. Besides, given the high cost of private hospitalisation, free treatment would have been desirable not only by the poor but also by the middle classes (see Chapter 4).

Contrary to their social physiognomy, pauper and well-off patients had very different medical profiles: psychiatric diagnoses were given only to a minority of paupers but to the majority of paying patients, while nervous and neurological diagnoses mainly concerned pauper patients (Table 12.4).[56] This divergence shows that the medical director's preference for neurological cases in free treatment was put into practice. The hospital's scientific interest in neurology, combined with the fact that most paying patients were psychiatric cases, and that overall fewer neurological and nervous patients were admitted to the Eginitio than psychiatric patients (see Table 8.3), explains why neurological and nervous cases were prioritised among paupers.

Table 12.1 Men and women among the pauper and well-off patients, Eginitio Hospital

Gender	*Pauper patients*		*"Well-off" patients*		*Patients of unknown class*	
	N	%	N	%	N	%
Men	131	61.5	738	61.1	6	66.7
Women	82	38.5	470	38.9	3	33.3
Total	213	100.0	1208	100.0	9	100.0

Source: Sample from the registers, Eginitio, 1905–1909 and 1916–1920 (see Table 1.1).

[54] Lorraine Walsh, "A class apart? Admissions to the Dundee Royal Lunatic Asylum, 1890–1910", in Jonathan Andrews and Anne Digby (eds), *Sex and Seclusion, Class and Custody: Perspectives on Gender and Class in the History of British and Irish Psychiatry*, Amsterdam: Rodopi, 2004, pp. 249–69.

[55] Anon., "Αιγινήτειον νοσοκομείον" ["The Eginitio hospital"], *Σκριπ [Skrip]*, 15/10/1904, p. 3.

[56] For the meanings of psychiatric, neurological and nervous diagnoses, see Chapter 8.

Table 12.2 Age groups among the pauper and well-off patients, Eginitio Hospital

Age group	*Pauper patients*		*"Well-off" patients*		*Patients of unknown class*	
	N	%	N	%	N	%
5–13	11	5.2	8	0.7	0	0.0
14–20	45	21.1	153	12.7	1	11.1
21–25	27	12.7	190	15.7	1	11.1
26–39	61	28.6	425	35.2	4	44.4
40–59	57	26.8	359	29.7	2	22.2
60+	9	4.2	69	5.7	1	11.1
Unknown	3	1.4	4	0.3	0	0.0
Total	213	100.0	1208	100.0	9	100.0

Source: Sample from the registers, Eginitio, 1905–1909 and 1916–1920 (see Table 1.1).

Table 12.3 Marital status of the pauper and well-off patients, Eginitio Hospital

Marital status	*Pauper patients*		*"Well-off" patients*		*Patients of unknown class*	
	N	%	N	%	N	%
Unmarried	124	58.2	625	51.7	6	66.7
Married	80	37.6	556	46.0	3	33.3
Widowed	6	2.8	26	2.2	0	0.0
Unknown	3	1.4	1	0.1	0	0.0
Total	213	100.0	1208	100.0	9	100.0

Source: Sample from the registers, Eginitio, 1905–1909 and 1916–1920 (see Table 1.1).

Table 12.4 Diagnoses of the pauper and well-off patients, Eginitio Hospital

Diagnoses	*Pauper patients*		*"Well-off" patients*		*Patients of unknown class*	
	N	%	N	%	N	5
Psychiatric	28	13.1	820	67.9	4	44.4
Neurological	137	64.3	272	22.5	5	55.6
Nervous	47	22.1	116	9.6	0	0.0
Unknown	1	0.5	0	0.0	0	0.0
Total	213	100.0	1208	100.0	9	100.0

Source: Sample from the registers, Eginitio, 1905–1909 and 1916–1920 (see Table 1.1).

The scientific purpose and function of free hospitalisation are confirmed by the length of hospitalisation of the paupers. Although, as shown in the previous chapter, the length of stay at the Eginitio was generally short, the paupers spent even shorter periods there. While the paying patients stayed in hospital for an average of 193 days, the average duration of hospitalisation for the paupers was almost half (93 days). This underlines that paupers were hospitalised less for treatment than for scientific observation and teaching. For students and medical staff to see many different cases, it was necessary that the few places for the paupers were not occupied by the same patients for long periods.

The characteristics of free and reduced-cost hospitalisation in the Eginitio, namely its scarcity, medical objectives, and short duration, excluded a wide range of patients, especially chronic and psychiatric ones. Although Katsaras had expected private patients to flock to the Eginitio,[57] he seemed not to have anticipated that many others would ask to be treated free of charge. The requests for free hospitalisation in the Eginitio testify to the need for free psychiatric care in Greece at the beginning of the twentieth century. The university hospital, however, was not founded to meet social needs, but to promote the scientific, academic, and professional organisation of Greek neurology and psychiatry. These aims of the Eginitio were negotiated, but not overturned, by wider social demands; free hospitalisation retained its primarily academic and scientific function, controlled by the hospital's medical authority.

The two hospitals studied were not only loci of institutional mental health care, but also loci of consolidation and reproduction of the psychiatric profession; each provided, to varying degrees and in different ways, opportunities for professional organisation, training, and research in psychiatry. The Eginitio, being a university hospital, was more oriented towards teaching and research than the Dromokaitio, and can also be seen as part of the efforts made in many countries since the end of the nineteenth century to improve the professional and social status and the public image of psychiatry, and to expand its field so that it could play a more dynamic public role and have access to more numerous and varied patients. Indeed, as shown in Part II of this book, the Eginitio seems to have been perceived differently and to have had a greater appeal to certain patients (children, women and nervous/neurological patients)

[57] "Minutes of the meeting of the rectorate, 11/12/1899".

who were less easily admitted to the other main psychiatric institution of the time, the Dromokaitio, which belonged to the psychiatric tradition of the asylums. It is not possible to know whether the alternative model of care offered by the Eginitio was beneficial for its patients, but it can be argued that the new university hospital, mainly because of its small capacity and its scientific and professional objectives, proved inadequate to meet the existing social needs for free mental health care with fewer procedures.

Epilogue

Focusing on the social reception and position of psychiatric institutions in early twentieth-century Greece, this book approached mental illness and psychiatric hospitalisation from a social and cultural perspective. It analysed the uses and functions of hospitals and the factors that determined how psychiatry and psychiatric institutions were approached. Furthermore, it explored the similarities and differences between medical and lay perceptions and practices of psychiatric hospitalisation, and the relationships and communication between institutions and their social environment. In these ways, the analysis attempted to move beyond an impersonal study of mental hospitals and to reach out to the many subjects involved in institutional mental health care—the patients and their relatives, friends and colleagues; the doctors and administrators of the hospitals; public officials and figures—and to trace the multifaceted experiences of mental illness and psychiatric hospitalisation.

Greece shared a common psychiatric background with the countries of the Western world, as evidenced by similarities in psychiatric legislation and medical discourse, the centrality of mental hospitals in mental health care and the important role of the medical director in these hospitals. However, Greece, a largely rural and traditional country, developed different formal responses to mental illness and different psychiatric services than the more urbanised and industrialised countries. Greek psychiatric law was vague and centralised, and mental hospitals were few,

D. Kritsotaki, *A Social History of Mental Health Care in Greece, c. 1900–1920*, Mental Health in Historical Perspective,
https://doi.org/10.1007/978-3-032-05764-8

concentrated in urban centres and operated on a national rather than local basis, combining private and public admissions, paid and free hospitalisation. The limited number of psychiatric institutions—especially public mental hospitals—and the complex way in which free mental health care was distributed made institutional mental health care inadequate and limited its integration into ways of dealing with mental illness.

Beyond the make-up of the psychiatric system and the wider social and economic conditions associated with its development and organisation, a variety of factors shaped the use of institutions. Historical circumstances certainly played a role: during the First World War and the Greco-Turkish War, the public use of institutions was expanded due to an increase of "service patients". Other factors determining how and by whom hospitals were used were the characteristics of each hospital—its aims and admission policy, the intensity of public supervision, the height of fees, its location, size and public image—and undoubtedly the health and socio-economic condition of the patients: on the one hand, patients with severe and chronic symptoms, violent, noisy and uncontrollable behaviour, who were not useful to their environment, did not work or did not take care of themselves, were more likely to be admitted to and stay longer periods in the institutions; on the other hand, admission and discharge were linked to the social and economic realities of patients' lives. The distance of home from the institution, although not a barrier, reduced the likelihood of admission and favoured long-term and continuous hospitalisation, while the financial resources of patients and their environment and the availability of extramural support, care, treatment, and supervision influenced the probability, speed, and duration of hospitalisation. A key factor was found to be the absence or presence of family members willing and able to care for the patients; the family was seen and did act as a major provider of support, care, and control.

In combination with the specific life and health circumstances of the patients, the extent and ways of the appropriation of institutional mental health care depended on social and cultural perceptions of mental illness and psychiatric hospitals. In general, the use of institutions was restricted and delayed by the non-recognition of mental illness as such or as a condition requiring psychiatric care; by tolerance towards the mentally ill; and by negative perceptions of institutions as harsh places that stigmatised patients and their families and were intended for "dangerous" and destitute patients. Moreover, attitudes towards mental illness and mental hospitals were linked to patients' social characteristics—gender, age, and social class—and to ideals and values about family, marriage, and work. Hospitalisation was a less acceptable and less likely resource for social groups perceived as more vulnerable and belonging to the private sphere, and less visible, controllable, and useful in the public sphere.

The role of the medical profession in the institutions was central, and medical theory and praxis had a special place among the concepts and practices that shaped institutional health care. Hospitalisation was heavily influenced by medical theories of mental illness and what was best for patients and society, but also by the scientific and professional interests of psychiatry, including the reproduction of the psychiatric profession and the enhancement of the professional and social status of doctors through teaching and research. The scientific and professional terms and purposes of institutional mental health care were not isolated from their social and cultural context. Rather, they communicated with lay concepts of who should be hospitalised. After all, as this book has shown, medical and lay perceptions of mental illness were largely based on shared social and cultural assumptions. As a result, medical and lay narratives overlapped in many respects, such as the causes of mental illness, the social characteristics of the mentally ill, and the situations that were considered appropriate for hospitalisation.

Nonetheless, there were points of divergence between medical and lay perceptions and practices. While the psychiatric profession sought to play an active role in (re)fashioning social reality, attempting to regulate social attitudes and cultural values, in practice psychiatrists in the early twentieth century had not assumed such a broad public role. Attempts to reform psychiatric hospitalisation through university clinics had brought less severe cases into institutional psychiatry, while medical theories and methods had to some extent infiltrated lay perceptions and practices, as evidenced by the fact that people with behaviours at odds with social norms were sent to the institutions—a practice that echoed the psychiatric assessment of such behaviours as symptoms of mental illness that could be treated in specialised hospitals. Nevertheless, admissions for mild disorders and behavioural problems were rare compared to admissions for serious physical and mental symptoms. This suggests that less severe disorders and conditions bordering on disruptive behaviour were either not considered to be mental illnesses or were thought to require a different type of psychiatric intervention from that provided in institutions, which was not available in Greece at the time.

Psychiatric aspirations were counterbalanced by the claims of several actors to the use of the mental hospitals. Medical control of institutional mental health care was not absolute: lay private and public actors were involved in the initiation and termination of hospitalisation, including hospital managers (from the private, public and university sectors), patients and their families, local and state authorities, various institutions and charitable and university agents. These actors were not all equally powerful or successful in their claims and did not always operate in the same way. Patients had less influence on hospitalisation decisions than their private and public environment and doctors; the role of private networks was more extensive than that of public networks; and the medical and administrative authority of institutions had more control over patients who were treated free of charge than over those who paid. Different actors often worked in combination, either in synergy or in opposition, and the action of one could limit that of the others. In any case, their dynamics were not fixed but fluid.

Therefore, the ways in which psychiatric hospitalisation was perceived and utilised varied. They were determined by processes of agreement, compromise, and conflict between the professional interests of doctors, the diverse and complex relationships of love, dependency and power in the private sphere, the need to maintain public order and to promote the charitable and political activities of public agents and individuals. For example, the perception of mental hospitals as places of permanent or unjust confinement in the context of family machinations or political persecutions, propagated by the press, conflicted with the image of mental hospitals as loci of care and treatment for the ill, an image projected by doctors. The latter argued that the condition of patients was assessed according to medical criteria that determined who should be admitted and who should be discharged. Conversely, non-medical actors often turned to hospitals, among other medical and lay means of treatment and relief, with the primary concern of solving their own problems and serving their own interests. But, although hospitals were claimed in different ways by different actors, their public and private functions were not strictly separated. The margins of action of public and private agents were not set but shifting, and hospitalisation, regardless of who mobilised it, had more or less the same objectives: care, treatment, rehabilitation, but also confinement, isolation, control and discipline of patients; temporary or long-term relief of the patients' environment from problems arising from their condition and behaviour.

By highlighting the diversity of actors and functions of psychiatric hospitalisation, this book has painted a contradictory picture of the social position of institutions and their patients. On the one side, psychiatric hospitals appear to have been embedded in various private and public practices and to have maintained multiple connections with their social surroundings. These relationships were formalised in psychiatric legislation and materialised through the ways in which hospitals were managed, financed and interacted with the outside world for example via visits of friends, relatives, authorities, and journalists. The hospitals' communication with their social milieu is particularly evidenced by the fact that hospitalisation, its initiation and duration, was adjusted to the needs, attitudes, values, and beliefs of the patients and their environment. Moreover, the relatively high rates of admission within a short time of the onset of illness, the predominantly short length of hospital stays, the practice of readmission and trial discharge suggest that entry into and exit from institutions were not too difficult. This meant that inmates and hospitals were not completely cut off from their social environment. On the other side, the reluctance and mistrust towards hospitalisation, the low penetration of psychiatric theories and practices into lay attitudes towards mental illness, the attempts at other forms of treatment and, above all, the high level of involuntary hospitalisation make the institutions seem isolated, with little and difficult contact and interaction with their social milieu.

The ambivalent social position of psychiatric institutions in early twentieth-century Greece is the main finding of this book, and it was rooted precisely in the variety and diversity of the factors that, as this research has shown, shaped the complex phenomena of mental illness and psychiatric hospitalisation. Inevitably, depending on the dynamics of these factors, psychiatric institutions—and with them the mentally ill—emerged both distant from and closely connected to their social environment.

Bibliography

Archival sources

Archive of the Eginitio Hospital, Athens

Α΄ Μητρώον Εσωτερικών Ασθενών Αιγινήτειου Νοσοκομείου [First Register of Inpatients of the Eginitio Hospital], 1.

Γ΄ Μητρώον Εσωτερικών Ασθενών Αιγινήτειου Νοσοκομείου [Third Register of Inpatients of the Eginitio Hospital], 1.1.

Πρόχειραι Ιστορίαι Νοσούντων Α΄ [Draft Histories of The Ill A΄], 9.1.

Πρόχειραι Ιστορίαι Νόσου Εισαγομένων Ασθενών Β΄ [Draft Illness Histories of Patients Admitted B΄], 9.2.

Historical Archive of the University of Athens, Athens

Κλινικές-Νοσοκομεία [Clinics-Hospitals], 1908–1909, 1909–1910, 1910–1911, 1911–1912, 1912–1913, 1913–1914, 1914–1915, 1915–1916, 1916–1917, 1917–1918, 1918–1919, 1919–1920, Archive of the Rectorate of the University of Athens.

Archive of the Dromokaitio Hospital, Athens

Βιβλίον των κατ' Αίτησιν Ιδιωτών Εισαγομένων Φρενοβλαβών (*Books of the Insane Admitted at the Request of Private Individuals]*, 1904–1906, 1906–1909, 1909–1911, 1915–197, 1917–1918, 1918–1919, 1919–1921.

D. Kritsotaki, *A Social History of Mental Health Care in Greece, c. 1900–1920*, Mental Health in Historical Perspective,
https://doi.org/10.1007/978-3-032-05764-8

Α΄ Βιβλίον των εξ Επαγγέλματος Εισαγομένων Ψυχοπαθών [First Book of ex Officio Admitted Psychopaths].

Β΄ Βιβλίον των εξ Επαγγέλματος Εισαγομένων Ψυχοπαθών [Second Book of ex Officio Admitted Psychopaths].

Βιβλίον Ιστορικών [Case histories book], 1900–1902, 1904–1907, 1907–1912, 1915–1916, 1916–1917, 1917, 1918, 1918–1919, 1919–1920, 1920, 1920–1921.

Published sources

"Περί εγκρίσεως Οργανισμού του Αιγινήτειου Νοσοκομείου" ["On the approval of the Statute of the Eginitio Hospital"], *Εφημερίς της Κυβερνήσεως [Government Gazette]*, 169, 30/7/1904.

"Περί εγκρίσεως του προσωρινού Κανονισμού του συσταθέντος Φρενοκομείου του αοιδίμου Ζωρζή Δρομοκαΐτου" ["On the approval of the provisional Regulation of the established Lunatic Asylum of the Venerable Zorzis Dromokaitis"], *Εφημερίς της Κυβερνήσεως [Government Gazette]*, 192, 16/7/1887.

Allbutt, Thomas Clifford, "Children, insanity in", in Daniel Hack Tuke (ed.), *A Dictionary of Psychological Medicine: Giving the Definition, Etymology and Synonyms of the Terms used in Medical Psychology with the Symptoms, Treatment, and Pathology of Insanity and the Law of Lunacy in Great Britain and Ireland*, v. 1, London: J. & A. Churchill, 1892, pp. 202–5.

André-Thomas, M., "Θεραπεία της υστερίας και της νευρασθενείας δια της απομονώσεως και της ψυχοθεραπείας" ["Treatment of hysteria and neurasthenia through isolation and psychotherapy"], *Ψυχιατρική και Νευρολογική Επιθεώρησις [Psychiatric and Neurological Review]*, 2 (1903–1904), pp. 336–48.

Anon., "Αι αποκαλύψεις του φρενοκομείου" ["The revelations of the lunatic asylum"], *Εμπρός [Forward]*, 16/10/1901, pp. 1–2.

Anon., "Αι αποκαλύψεις του φρενοκομείου" ["The revelations of the lunatic asylum"], *Εμπρός [Forward]*, 18/10/1901, pp. 1–2.

Anon., "Αι οφειλαί του Δήμου προς το Δρομοκαΐτειον" ["The debts of the Municipality to the Dromokaitio"], *Σκριπ [Skrip]*, 11/5/1904, p. 2.

Anon., "Αι πανεπιστημιακαί κλινικαί και οι φοιτηταί της Ιατρικής" ["University clinics and medical students"], *Σκριπ [Skrip]*, 5/12/1909, p. 2.

Anon., "Αι σκηναί του Δρομοκαΐτείου" ["The events in the Dromokaitio"], *Εμπρός [Forward]*, 9/4/1902, p. 2.

Anon., "Αι χθεσιναί σκηναί εις το Δρομοκαΐτειον" ["Yesterday's events in the Dromokaitio"], *Εμπρός [Forward]*, 8/4/1902, p. 2.

Anon., "Αιγινήτειον νοσοκομείον" ["The Eginitio Hospital"], *Σκριπ [Skrip]*, 15/10/1904, p. 3.

Anon., "Διάγνωσις της νευρασθενείας" ["Diagnosis of neurasthenia"], *Ιατρική Πρόοδος [Medical Progress]*, 20 (1915), pp. 95–96.

Anon., "Ειδήσεις: Το φρενοκομείο της Κερκύρας" ["News: The lunatic asylum of Corfu"], *Ψυχιατρική και Νευρολογική Επιθεώρησις [Psychiatric and Neurological Review]*, 6 (1903), p. 192.

Anon., "Επιθεώρησις εις το Δρομοκαΐτειον" ["Inspection at the Dromokaitio"], *Εμπρός [Forward]*, 20/6/1902, p. 2.

Anon., "Η διπλή Βουλή. Επιψήφισις νομοσχεδίων" ["The double Parliament. Voting on bills"], *Σκριπ [Skrip]*, 18/2/1911, p. 3.

Anon., "Η εκπαίδευσις των κορασίων και η κληρονομικότης (κατά τον Γκυγιώ)" ["Girls' education and heredity (according to Guyau)"] (1901), in Eleni Fournaraki, *Εκπαίδευση και Αγωγή των Κοριτσιών: Ελληνικοί Προβληματισμοί (1830–1910): Ένα Ανθολόγιο [Schooling and Education of Girls: Greek Reflections (1830–1910): An Anthology]*, Athens: Γενική Γραμματεία Νέας Γενιάς [General Secretariat for Youth], 1987, pp. 483–86.

Anon., "Η επίσκεψις του Νομάρχου εις το Δρομοκαΐτειον" ["The Prefect's visit to Dromokaitio"], *Εμπρός [Forward]*, 18/10/1901, p. 2.

Anon., "Η χθεσινή συνεδρίασις του δημ. συμβουλίου Πειραιώς" ["Yesterday's meeting of the Piraeus City Council"], *Εμπρός [Forward]*, 16/3/1908, p. 4.

Anon., "Θεραπευτήριον ο 'Ευαγγελισμός'" ["The 'Evangelismos' Sanatorium"], *Εμπρός [Forward]*, 16/2/1915, p. 2.

Anon., "Ο δραπέτης του Δρομοκαϊτείου" ["The escapee of the Dromokaitio"], *Εμπρός [Forward]*, 15/10/1901, p. 2.

Anon., "Ο κ. Νομάρχης εις το Δρομοκαΐτειον και το Εφηβείον" ["The Prefect at the Dromokaitio and the Adolescents' Reformatory"], *Εμπρός [Forward]*, 5/2/1902, p. 3.

Anon., "Τα του φρενοκομείου" ["Issues of the lunatic asylum"], *Εμπρός [Forward]*, 18/10/1901, p. 1.

Anon., "Το Δρομοκαΐτειον γεμάτο!" ["The Dromokaitio full!"], *Εμπρός [Forward]*, 30/6/1902, p. 3.

Anon., "Το σκάνδαλον των οφειλών του δήμου εις το Δρομοκαΐτειον" ["The scandal of the municipality's debts to Dromokaitio"], *Σκριπ [Skrip]*, 13/5/1904, p. 4.

Anon., "Το χθεσινόν δημοτικόν συμβούλιον" ["Yesterday's city council meeting"], *Σκριπ [Skrip]*, 11/3/1905, p. 1.

Apostolidis, Simon, "Ο αλκοολισμός εν Κωνσταντινοπόλει παρά τοις ομογενέσι κατοίκοις" ["Alcoholism in Constantinople among the Greek community"], *Ψυχιατρική και Νευρολογική Επιθεώρησις [Psychiatric and Neurological Review]*, 6 (1907–1908), pp. 275–84, 335–43.

Apostolidis, Simon, *Αι Ψυχώσεις [The Psychoses]*, Athens: s.n., 1889.

Arkalidis, Nikolaos, "Οι τοξικομανείς εν Ελλάδι" ["Drug addicts in Greece"], in *Πανηγυρικόν Τεύχος επί τη Τριακονταπενταετηρίδι Μιχαήλ Κατσαρά: Εκδοθέν υπό*

Συναδέλφων και Μαθητών αυτού, 29 Μαρτίου 1928 [Commemorative Issue on the Thirty-fifth Anniversary of Michael Katsaras: Published by his Colleagues and Students, 29 March 1928], Athens: Π. Δ. Σακελλάριος [P.D. Sakellarios Publishing House], 1928, pp. 91–97.

Armodios, Achilleas, "Ο αλκοολισμός υπό υγιεινήν και κοινωνιολογικήν έποψιν" ["Alcoholism from a hygienic and sociological point of view"], *Ψυχιατρική και Νευρολογική Επιθεώρησις [Psychiatric and Neurological Review]*, 2 (1903–1904), pp. 14–23, 40–48, 71–80.

Asylum for the Incurable, *Άσυλο των Ανιάτων υπό την Προστασίαν της Α.Μ. της Βασιλίσσης, Έκθεσις των Πεπραγμένων του έτους 1907 [Asylum for the Incurable under the Protection of her Majesty the Queen, Report of the Activities of the year 1907]*, Athens: s.n., 1908.

Asylum for the Incurable, *Άσυλον Ανιάτων, Καταστατικόν 1914 [Asylum for the Incurable, Statute, 1914]*, Athens: s.n., 1915.

Beach, Fletcher, "Insanity in children", *Journal of Mental Science*, 44 (1898), pp. 459–73.

Beach, Fletcher, *The Treatment and Education of Mentally Feeble Children*, London: J. & A. Churchill, 1895.

Breuer, Joseph, "Theoretical issues" (1895), in Joseph Breuer and Sigmund Freud, *Studies in Hysteria*, London: Penguin, 2004, pp. 189-254.

Bucknill, John Charles and Daniel Hack Tuke, *A Manual of Psychological Medicine: Containing the History, Nosology, Description, Statistics, Diagnosis, Pathology, and Treatment of Insanity: with an Appendix of Cases*, London: J. Churchill, 1858.

Burrows, George Man, "Commnetaries on insanity" (1828), in Vieda Skultans, *Madness and Morals: Ideas on Insanity in the Nineteenth Century*, London: Routledge & Kegan Paul, 1975, pp. 223–27.

Chaslin, Philippe, *La Confusion Mentale Primitive, Stupidité, Démence Aiguë, Stupeur Primitive*, Paris: Asselin et Houzeau, 1895.

Clifford Allbutt, Thomas and Walter Ernest Dixon, "Opium poisoning and other intoxicants", in Thomas Clifford Allbutt and Humphry Davy Rolleston (eds), *A System of Medicine by Many Writers*, London, 1906, pp. 965–68.

Clouston, Thomas, "Developmental insanities and psychoses. The delirium and night terrors, &c., of children. The insanities of children. The insanities of puberty and adolescence", in Daniel Hack Tuke (ed.), *A Dictionary of Psychological Medicine: Giving the Definition, Etymology and Synonyms of the Terms used in Medical Psychology with the Symptoms, Treatment, and Pathology of Insanity and the Law of Lunacy in Great Britain and Ireland*, v. 1, London: J. & A. Churchill, 1892, pp. 357–71.

Clouston, Thomas, *Clinical Lectures on Mental Diseases*, 5th edition, London: J. Churchill, 1898.

Clouston, Thomas, *The Hygiene of Mind*, London: Methuen & Co, 1906.

Craig, Maurice, *Psychological Medicine. A Manual of Mental Diseases for Practitioners and Students*, 3rd edition, London: J. Churchill, 1917.

de La Tourette, Gilles, *Les États Neurasthéniques: Formes Cliniques, Diagnostic, Traitement*, Paris: J.-B. Baillière et fils, 1898.

Déjerine, Jules, "Μέθοδος της εν νοσοκομείω απομονώσεως δια τας νευρώσεις" ["Method of hospital isolation for neuroses"], *Ψυχιατρική και Νευρολογική Επιθεώρησις [Psychiatric and Neurological Review]*, 2 (1903–1904), pp. 288–89.

Donkin, Horatio Bryan, "Hysteria" (1892), in Sally Ledger and Roger Luckhurst (eds), *The Fin de Siècle: A Reader in Cultural History, c.1880–1900*, Oxford: Oxford University Press, 2000, pp. 245–50.

Down, John Langdon, *On Some of the Mental Affections of Childhood and Youth*, London: J. & A. Churchill, 1887.

Dromokaitio Hospital, *Απολογισμός και Στατιστική του Δρομοκαϊτείου Φρενοκομείου Έτους 1889 [Report and Statistic of the Dromokaitio Lunatic Asylum for the year 1889]*, Athens: s.n., 1890.

Dubois, Paul, "Αι ψυχονευρώσεις και η ηθική αυτών θεραπεία" ["Psychoneuroses and their moral treatment"] *Ψυχιατρική και Νευρολογική Επιθεώρησις [Psychiatric and Neurological Review]*, 2 (1903–1904), pp. 348–56.

Durkheim, Émile, *L'Education Morale*, Paris: Presses Universitaires de France, 1974 (first edition in 1925).

Eager, R., "A record of admissions to the mental section of the Lord Derby War Hospital, Warrington, from June 1916 to June 16th 1917", *Journal of Mental Science*, 64 (1918), pp. 272–96.

Economakis, Miltiadis, "Η πρωτόγονος άνοια (dementia praecox). Κλινική περιγραφή εν συνοδεία ιδίων παρατηρήσεων" ["Dementia praecox. Clinical description accompanied by own observations"] (1906), in Athanasios Karavatos, Dimitris Ploumpidis and Yoryos Christodoulou (eds), *Ανθολόγιο Ελληνικών Ψυχιατρικών Κειμένων [Anthology of Greek Psychiatric Texts]*, Athens: Ελληνική Ψυχιατρική Εταιρεία [Hellenic Psychiatric Association], 2006, pp. 239–49.

Ellis, Havelock, "Sex, influence of, in insanity", in Daniel Hack Tuke (ed.), *A Dictionary of Psychological Medicine: Giving the Definition, Etymology and Synonyms of the Terms used in Medical Psychology with the Symptoms, Treatment, and Pathology of Insanity and the Law of Lunacy in Great Britain and Ireland*, v. 2, London: J. & A. Churchill, 1892, pp. 1152–56.

Esquirol, Jean-Étienne Dominique, *Des Maladies Mentales: Considérées sous les Rapports Médical, Hygiénique et Médico-Légal*, 2 vols, Paris: J.-B. Baillière, 1858.

Hall, Granville Stanley, *Adolescence: Its Psychology and Its Relations to Physiology, Anthropology, Sociology, Sex, Crime, Religion, and Education*, v. 2, London: S. Appleton, 1904.

Hammond, William, *A Treatise on Insanity in its Medical Relations*, New York, 1883.

Ireland, William, *The Mental Affections of Children: Idiocy, Imbecility and Insanity*, 2nd edition, London: J. & A. Churchill, 1900 (first edition in 1898).

Kahlbaum, Karl, "On heboidophrenia" (1890), *History of Psychiatry*, 13 (2002), pp. 201–8.

Katsaras, Michael, "Περί ψευδών κατατονικών ανοιών. Λόγος εκφωνηθείς τη 20ή Ιανουαρίου 1908 εν τη μεγάλη αιθούση του πανεπιστημίου" ["On pseudo catatonic dementias. Speech delivered on 20 January 1908 in the great hall of the university"], in National University, *Τα κατά την Πρυτανείαν Μιχαήλ Κατσαρά [The Activities of the Rectorate of Michael Katsaras]*, Athens: Π. Δ. Σακελλαρίου [P.D. Sakellarios Publishing House], 1909, pp. 4–6.

Katsaras, Michael, *Παθολογία των Νεύρων και Ψυχιατρική. Νευρώσεις, Νευρίτιδες, Περιφερικαί Παραλύσεις [Pathology of the Nerves and Psychiatry. Neuroses, Diseases of the Nerves, Peripheral Paralyses]*, v. 3, Athens: Εστία [Estia], 1900.

Katsaras, Michael, *Παθολογία των Νεύρων και Ψυχιατρική. Νόσοι του Εγκεφάλου, Νόσοι της Παρεγκεφαλίδος, Φρενίτιδες Νόσοι [Pathology of the Nerves and Psychiatry. Diseases of the Brain, Diseases of the Cerebellum, Phrenic Diseases]*, v. 2, Athens: Εστία [Estia], 1898.

Kourtidis, Aristotle, "Ψυχοπάθειαι εν τω σχολείω. Νύξεις προς τους διδασκάλους και τους γονείς" ["Psychopathies in the school. Suggestions to teachers and parents"], *Ψυχιατρική και Νευρολογική Επιθεώρησις [Psychiatric and Neurological Review]*, 2 (1903–1904), pp. 81–89.

Kouzis, Aristotle, *Εθνικό και Καποδιστριακό Πανεπιστήμιο Αθηνών, Εκατοενταετηρίς 1837–1937, Γ΄ Ιστορία της Ιατρικής Σχολής [National and Kapodistrian University of Athens, Centenary 1837–1937, v. 3 History of the Medical School]*, Athens: Τύποις "Πυρσού" [Pyrsos Publishing House], 1939.

Kraepelin, Emil, "Lectures on clinical psychiatry" (1894), in Roy Porter (ed.), *The Faber Book of Madness*, London: Faber and Faber, 1991, pp. 393–97.

Kraepelin, Emil, *Cent Ans de Psychiatrie, Suivi de La Folie Maniaco-Depressive* (1917), transl. Marc Geraud, Bordeaux: Mollat, 1997.

Libert, Lucien, "Les aliénés en Orient (Grèce, Empire Ottoman, Égypte)", *L'Informateur des Aliénistes et des Neurologistes*, Supplement de l' *Encéphale*, 7 (1912) pp. 283–302.

Magnan, Valenin and Paul Légrain, *Les Dégénérés. État Mentale et Syndromes Episodiques*, Paris: Rueff, 1895.

Magnan, Valenin, *Leçons Cliniques sur les Maladies Mentales*, Paris: Publications de Progrès Medical, 1891.

Magnan, Valentin and Paul Légrain, "Delusion in degenerates. The immediate delusions", in Francois-Regis Cousin, Jean Garrabé and Denis Morozou (eds),

Anthology of French Language Psychiatric Texts, transl. John Crisp, Le Plessis-Robinson: Institut Synthélabo pour le Progrès de la Connaissance, 1999, pp. 267–83.

Magnan, Valentin, *On Alcoholism: The Various Forms of Alcoholic Delirium and their Treatment*, transl. W. S. Greenfield, London: H. K. Lewis, 1876.

Makris, Ilias, "Περίπτωσις υστερικών εγκαυμάτων" ["Case of hysterical burns"], *Ιατρική Πρόοδος [Medical Progress]*, 23 (1918), pp. 147–52.

Makris, Ilias, *Οι Φρενοπαθείς Εγκληματίαι Αλλαχού και εν Ελλάδι [The Insane Criminals in Greece and Elsewhere]*, Athens: s.n., 1924.

Maudsley, Henry, "Body and mind: An inquiry into their connection and mutual influence, specially in reference to mental disorders" (1873), in Vieda Skultans, *Madness and Morals: Ideas on Insanity in the Nineteenth Century*, London: Routledge & Kegan Paul, 1975, pp. 230–35.

Maudsley, Henry, *The Pathology of Mind: A Study of its Distempers, Deformities, and Disorders*, London: Macmillan, 1995 (first published in 1895).

Maudsley, Henry, *The Physiology and Pathology of the Mind*, London: Routledge/ Thoemmes Press, 1993 (first edition in 1867).

Mercier, Charles Arthur, *A Text-book of Insanity and Other Mental Diseases*, London: George Allen and Unwin, 1914.

Ministry of the Interior, *Δημοσιεύματα του Τμήματος της Δημοσίας Οικονομίας [Publications of the Department of Public Finance]*, no. 4, Athens: s.n., 1899.

Ministry of the Interior, *Στατιστική του εν Κερκύρα Φρενοκομείου του Έτους 1877 [Statistic of the Lunatic Asylum in Corfu in the Year 1877]*, Athens: Εθνικό Τυπογραφείο [National Press], 1878.

Mitaftsis, Konstantinos, *Ψυχιατρική μετά 54 Προσωπογραφιών [Psychiatry with 54 Portraits]*, Athens: s.n., 1933.

Moreau (de Tours), Paul, *La Folie chez les Enfants*, Paris: J.-B. Baillière, 1888.

Morel, Bénédict-Augustin, "Treatise on mental diseases", in Francois-Regis Cousin, Jean Garrabé and Denis Morozou (eds), *Anthology of French Language Psychiatric Texts*, transl. John Crisp, Le Plessis-Robinson: Institut Synthélabo pour le Progrès de la Connaissance, 1999, pp. 179–80.

Morel, Bénédict-Augustin, *Traité des Dégénéréscences Physiques, Intellectuelles et Morales de l'Éspece Humaine et des Causes qui produisent ces Variétés Maladives*, v. 1, Paris: J.-B. Baillière, 1857.

Morison, Alexander, "Outlines of lectures on mental diseases (1824)", in Vieda Skultans, *Madness and Morals: Ideas on Insanity in the Nineteenth Century*, London: Routledge & Kegan Paul, 1975, pp. 32–33.

National University, *Τα κατά την Πρυτανείαν Γεωργίου Ν. Χατζιδάκι, Τακτικού Καθηγητού της Γλωσσολογίας, Πρυτανεύσαντος κατά το Ακαδημαϊκόν Έτος 1905–1906 [The Activities of the Rectorate of Yeoryios N. Hadjidakis, Professor of Linguistics, Rector during the Academic Year 1905–1906]*, Athens: Τύποις Π.Δ. Σακελλαρίου [P.D. Sakellarios Publishing House], 1907.

National University, *Τα κατά την Πρυτανείαν Σπυρ. Π. Λάμπρου Πρυτανεύσαντος κατά το Ακαδημαϊκόν Έτος 1904–1905 [The Activities of the Rectorate of Spyr. P. Lambros Rector during the Academic Year 1904–1905]*, Athens: Τύποις Π.Δ. Σακελλαρίου [P.D. Sakellarios Publishing House], 1907.

Papastratiyakis, Konstantinos, "Τα νευρικά και ψυχικά υπολείμματα των τραυματισμών του κρανίου" ["The nervous and mental remnants of skull injuries"], *Ιατρική Πρόοδος [Medical Progress]*, 14 (1927), pp. 369–80.

Price, G., "Αιτιολογία των μη οργανικών φρενοπαθειών" ["Aetiology of the non-organic phrenopathies"], *Ιατρική Πρόοδος [Medical Progress]*, 17 (1912), pp. 55–58.

Ray, M. B., "Case of acute mania occurring in a boy", *Journal of Mental Science*, 44 (1898), pp. 320–21.

Régis, Emmanuel, *Précis de Psychiatrie*, 5th edition, Paris: Octave Doin et fils, 1914.

Tuke, Daniel Hack, *Chapters in the History of the Insane in the British Isles*, London: Kegan Paul, Trench, 1882.

University of Athens, *Επετηρίδες των Πανεπιστημιακών Ετών 1864–1865 – 1892–1893 [Yearbooks of the Academic Years 1864-1865–1892-1893]*, Athens, 1865–1895.

Vlavianos, Simonidis, "Διαφορική διάγνωσις νευρασθενείας και γενικής παραλύσεως μετά πρωτοτύπων κλινικών παρατηρήσεων" ["Differential diagnosis of neurasthenia and general paralysis with original clinical observations"], *Ψυχιατρική και Νευρολογική Επιθεώρησις [Psychiatric and Neurological Review]*, 2 (1903–1904), pp. 132–35, 165–69, 199–204, 356–60.

Vlavianos, Simonidis, "Εθνική νευρασθένεια" ["National neurasthenia"], *Ψυχιατρική και Νευρολογική Επιθεώρησις [Psychiatric and Neurological Review]*, 6 (1907–1908), pp. 3–5.

Vlavianos, Simonidis, "Η διανοητική πρωτοπαθής σύγχυσις, καθορισμός και συνωνυμία" ["Mental primary confusion, definition and synonymy"], *Ψυχιατρική και Νευρολογική Επιθεώρησις [Psychiatric and Neurological Review]*, 3 (1904–1905), pp. 43–48.

Vlavianos, Simonidis, "Η ψυχοθεραπεία και η επίδρασις αυτής επί των ψυχονευρώσεων" ["Psychotherapy and its effect on psychoneuroses"], *Ψυχιατρική και Νευρολογική Επιθεώρησις [Psychiatric and Neurological Review]*, 2 (1903–1904), p. 321–25.

Vlavianos, Simonidis, "Ο αλκοολισμός και αι ολέθριαι αυτού συνέπειαι" ["Alcoholism and its disastrous consequences"], *Ψυχιατρική και Νευρολογική Επιθεώρησις [Psychiatric and Neurological Review]*, 6 (1907–1908), pp. 97–128.

Vlavianos, Simonidis, "Ο αλκοολισμός" ["Alcoholism"], *Ψυχιατρική και Νευρολογική Επιθεώρησις [Psychiatric and Neurological Review]*, 8 (1909–1910), pp. 3–192.

Vlavianos, Simonidis, "Ο ανταλκοολικός αγών εν Ελλάδι" ["The anti-alcoholic campaign in Greece"], *Ψυχιατρική και Νευρολογική Επιθεώρησις [Psychiatric and Neurological Review]*, 8 (1909–1910), pp. 1–2.

Vlavianos, Simonidis, "Περί των εκ των οινοπνευματωδών παθήσεων εν Ελλάδι" ["On the diseases caused by alcohol in Greece"], *Ψυχιατρική και Νευρολογική Επιθεώρησις [Psychiatric and Neurological Review]*, 1 (1902–1903), pp. 333–36.

Vlavianos, Simonidis, "Τα μη κανονικά παιδία" ["The non-normal children"], *Ψυχιατρική και Νευρολογική Επιθεώρησις [Psychiatric and Neurological Review]*, 6 (1907–1908), pp. 11–13.

Warnock, John, "Insanity from hasheesh", *Journal of Mental Science*, 49 (1903), pp. 96–110.

Yanniris, Michael and Nikolaos Arkalidis, "Η δι' ελονοσίας θεραπεία της προϊούσης καθολικής παραλύσεως" ["Malarial treatment of general paralysis of the insane"], in *Πανηγυρικόν Τεύχος επί τη Τριακονταπενταετηρίδι Μιχαήλ Κατσαρά: Εκδοθέν υπό Συναδέλφων και Μαθητών αυτού, 29 Μαρτίου 1928 [Commemorative Issue on the Thirty-fifth Anniversary of Michael Katsaras: Published by his Colleagues and Students, 29 March 1928]*, Athens: Π. Δ. Σακελλάριος [P.D. Sakellarios Publishing House], 1928, pp. 161–79.

Yanniris, Michael, "Η σύφιλις και αι ψυχοπάθειαι και η διαγνωστική αξία των βιολογικών ερευνών" ["Syphilis and psychopathies and the diagnostic value of biological investigations"], in *Πανηγυρικόν Τεύχος επί τη Τριακονταπενταετηρίδι Μιχαήλ Κατσαρά: Εκδοθέν υπό Συναδέλφων και Μαθητών αυτού, 29 Μαρτίου 1928 [Commemorative Issue on the Thirty-fifth Anniversary of Michael Katsaras: Published by his Colleagues and Students, 29 March 1928]*, Athens: Π. Δ. Σακελλάριος [P.D. Sakellarios Publishing House], 1928, pp. 83–87.

Yanniris, Michael, "Περί των εξ οινοπνεύματος παθήσεων εν Ελλάδι", *Ιατρική Πρόοδος*, 8 (1903), pp. 179–83; Simonidis Vlavianos, "Ο αλκοολισμός", *Ψυχιατρική και Νευρολογική Επιθεώρησις*, 8: 5–6 (1909–1910), pp. 3–192.

Yanniris, Michael, "Περί των εξ οινοπνεύματος παθήσεων εν Ελλάδι" ["On diseases caused by alcohol in Greece"], *Ιατρική Πρόοδος [Medical Progress]*, 8 (1903), pp. 179–83.

Yanniris, Michael, *Αι Φρενοπάθειαι εν Ελλάδι: Μελέτη επί Στατιστικών και Κλινικών Παρατηρήσεων [The Phrenopathies in Greece: A Study on Statistical and Clinical Observations]*, Syros: s.n., 1898.

Yanniris, Michael, *Περί Προϊούσης Καθολικής Παραλύσεως: Μελέτη Κλινική και Ιατροδικαστική [On General Paralysis of the Insane: Clinical and Forensic Study]*, Athens: s.n, 1900.

Yanniris, Michael, *Περί της Νοσηλείας και Περιθάλψεως των Φρενοπαθών εν τοις Φρενιατρείοις ή κατ' Οίκον: Προς Μόρφωσιν Νοσοκόμων [On the Care and Treatment of Insane Persons in Lunatic Asylums or at Home: For the Education of Nurses]*, Athens: s.n., 1903.

Zilanakis, Yeoryios, "Η προϊούσα καθολική παράλυσις. Στατιστική της δεκαπενταετίας 1908–1922" ["General paralysis of the insane. Statistic of the fifteen years 1908–1922"], *Ιατρική Πρόοδος [Medical Progress]*, 30 (1925), pp. 204–7.

Zografidis, S., "Η ψυχική κατάστασις των νευρασθενικών" ["The mental state of the neurasthenics"], *Ιατρική Πρόοδος [Medical Progress]*, 5 (1900), p. 303.

Απογραφή του Πληθυσμού της Ελλάδος κατά την 19 Δεκεμβρίου 1920. Γενικά Στατιστικά Αποτελέσματα [Census of the Population of Greece on 19 December 1920. General Statistical Results], Athens: Εθνικό Τυπογραφείο [National Press], 1928.

Λογοδοσία του Διοικητικού Συμβουλίου του εν Δαφνίω Θεραπευτηρίου Ζωρζή και Ταρσής Δρομοκαΐτου κατά το Έτος 1921 [Report of the Board of Directors of the Hospital of Zorzis and Tarsi Dromokaitis in Dafni during the Year 1921], Athens: s.n., 1922.

Λογοδοσία του Διοικητικού Συμβουλίου του εν Δαφνίω Φρενοκομείου Δρομοκαϊτείου, 1898: Πεπραγμένα [Report of the Board of Directors of the Dromokaitio Lunatic Asylum in Daphni, 1898: Activities], Athens: s.n., 1899.

Νόμος ΑΦΙΔ' "Περί αδείας συστάσεως ιδιωτικού φρενοκομείου εκ του κληροδοτήματος Δρομοκαΐτου" [Law 1514 "On the licence for the establishment of a private lunatic asylum from the Dromokaitis bequest"], *Εφημερίς της Κυβερνήσεως [Government Gazette]*, 145, 10/6/1887.

Νόμος ΓΩΚΓ' "Περί Οργανισμού του Εθνικού Πανεπιστημίου" [Law 3823 "On the Statute of the National University"], *Εφημερίδα της Κυβερνήσεως [Government Gazette]*, 178, 12/7/1911.

Νόμος ΨΜΒ' "Περί συστάσεως φρενοκομείων" [Law 742 "On the establishment of lunatic asylums", *Εφημερίς της Κυβερνήσεως [Government Gazette]*, 28, 19/5/1862.

Στατιστικά Αποτελέσματα της Γενικής Απογραφής του Πληθυσμού κατά την 27 Οκτωβρίου 1907 [Statistical Results of the General Census of the Population on 27 October 1907], v. 1, Athens: Εθνικό Τυπογραφείο [National Press], 1909.

Secondary Literature

Abi-Rached, Joelle, *Aṣfūriyyeh: A History of Madness, Modernity, and War in the Middle East*, Cambridge, MA: MIT Press, 2020.

Adair, Richard, Bill Forsythe and Joseph Melling, "A danger to the public? Disposing of pauper lunatics in late-Victorian and Edwardian England: Plympton St. Mary Union and the Devon County Asylum, 1867-1914", *Medical History*, 42 (1998), pp. 1–25.

Agriantoni, Christina, "Βιομηχανία" ["Industry"], in Christos Hadziiossif (ed.), *Ιστορία της Ελλάδας του 20ού Αιώνα [History of Greece in the 20th Century]*, v. A1, Athens: Βιβλιόραμα [Vivliorama], 1998, pp. 173-221.

Agriantoni, Christina, "Η ελληνική οικονομία. Η συγκρότηση του ελληνικού καπιταλισμού, 1870–1909" ["The Greek economy. The constitution of Greek capitalism, 1870–1909"], in Vasilis Panayotopoulos (ed.), *Ιστορία του Νέου Ελληνισμού, 1770–2000 [History of Modern Hellenism, 1770–2000]*, v. 5, Athens: Ελληνικά Γράμματα [Ellinika Grammata], 2003, pp. 55–70.

Agriantoni, Christina, *Οι Απαρχές της Εκβιομηχάνισης στην Ελλάδα τον 19ο Αιώνα [The Beginnings of Industrialisation in Greece in the 19th Century]*, Athens: Εμπορική Τράπεζα της Ελλάδος [Commercial Bank of Greece], 1986.

Ailianos, Michael, *Το Έργον της Ελληνικής Περιθάλψεως [The Work of Greek Welfare]*, Athens: s.n., 1921.

Alexander, Franz and Sheldon Selesnick, *The History of Psychiatry: An Evaluation of Psychiatric Thought and Practice from Prehistoric Times to the Present*, London: Allen & Unwin, 1967.

Alston, Margaret and Jenny Kent, "The big dry: The link between rural masculinities and poor health outcomes for farming men", *Journal of Sociology*, 44(2), 2008, pp. 133–47.

Ampatzoglou, Grigoris, "Αυτοβιογραφία, μνήμη, θεραπεία" ["Autobiography, memory, therapy"], in Rika Benveniste and Theodoros Paradellis (eds), *Διαδρομές και Τόποι της Μνήμης: Ιστορικές και Ανθρωπολογικές Προσεγγίσεις [Routes and Places of Memory: Historical and Anthropological Approaches]*, Athens: Αλεξάνδρεια [Alexandria], 1999, pp. 65-87.

Ampatzoglou, Grigoris, *Ψυχιατρική και Ιατρική. Τα Όρια, οι Ταυτότητες και οι Σχέσεις [Psychiatry and Medicine. The Boundaries, the Identities and the Relationships]*, Athens: Οδυσσέας [Odysseas], 1991.

Andrews, Jonathan, "Case notes, case histories and the patients' experience of insanity at Gartnavel Royal Asylum, Glasgow, in the nineteenth century", *Social History of Medicine*, 2 (1998), pp. 255-81.

Andriotis, Nikos, "Η πρώτη προσφυγιά. Ελληνικές προσφυγικές μετακινήσεις, 1906–1922" ["The first refugees. Greek refugee movements, 1906–1922"], in Vasilis Panayotopoulos (ed.), *Ιστορία του Νέου Ελληνισμού, 1770–2000 [History of Modern Hellenism, 1770–2000]*, v. 6, Athens: Ελληνικά Γράμματα [Ellinika Grammata], 2003, pp. 95–104.

Ankele, Monika and Benoît Majerus (eds), *Material Cultures of Psychiatry*, Bielefeld: Transcript, 2020.

Anon, *Τα 80 Χρόνια του Ασύλου Ανιάτων: 1893–1973 [The 80 years of the Asylum for the Incurable: 1893–1973]*, Athens: s.n., s.d., pp. 13–15).

Apesoa-Varano, Ester Carolina, Judith C. Barker and Ladson Hinton, "'If you were like me, you would consider it too': Suicide, older men, and masculinity", *Society and Mental Health*, 8(2), 2018, pp. 157–73.

Ariès, Philippe, *Centuries of Childhood*, New York: Vintage Books, 1960.

Armstrong, Walter Alan, "The use of information about occupation", in Edward Anthony Wrigley (ed.), *Nineteenth-Century Society: Essays in the Use of Quantitative Methods for the Study of Social Data*, London: Cambridge University Press, 1972, pp. 191–310.

Atzina, Lena, *Η Μακρά Εισαγωγή της Ψυχανάλυσης στην Ελλάδα [The Long Introduction of Psychoanalysis in Greece]*, Athens: Εξάντας [Exandas], 2004.

Avdela, Efi, "Οι γυναίκες, κοινωνικό ζήτημα" ["Women, a social issue"], in Christos Hadziiossif (ed.), *Ιστορία της Ελλάδας του 20ου αιώνα 1922-1940, Ο Μεσοπόλεμος [History of Greece in the 20th Century1922-1940, The Interwar Period]*, v. B1, Athens: Βιβλιόραμα [Vivliorama], 2002, pp. 337-59.

Avdela, Efi, *Δημόσιοι Υπάλληλοι Γένους Θηλυκού. Καταμερισμός της Εργασίας κατά Φύλα στον Δημόσιο Τομέα, 1908–1955 [Female Public Servants: Labour Division by Gender in the Public Sector, 1908–1955]*, Athens: Ίδρυμα Έρευνας και Παιδείας της Εμπορικής Τράπεζας της Ελλάδος [Research and Education Foundation of the Commercial Bank of Greece], 1990.

Baethge, Christopher, Ira Glovinsky and Ross Baldessarini, "Manic-depressive illness in children: An early twentieth-century view by Theodor Ziehen (1862-1950)", *History of Psychiatry*, 15 (2004), pp. 201–26.

Baethge, Christopher, Paola Samvatore and Ross Baldessarini, "Ewald Hecker, cyclothymia, a circular mood disorder", *History of Psychiatry*, 14 (2003), pp. 377–90.

Bajeux, Camille and Aude Fauvel, "'The nervous testicle'. Urology and psychiatry in Belle Époque France", *Histoire, Médecine et Santé*, 23 (2023), pp. 35–54.

Ballet, Gilbert, *Psychoses et Affections Nerveuses*, Paris: O. Doin, 1897.

Barmpouti, Alexandra, *Post-War Eugenics, Reproductive Choices and Population Policies in Greece 1950s-1980s*, Cham, Switzerland: Palgrave Macmillan, 2019.

Bartlett, Peter and David Wright, "Community care and its antecedents", in Peter Bartlett and David Wright (eds), *Outside the Walls of the Asylum: On "Care and Community" in Modern Britain and Ireland*, New Brunswick: Athlone Press, 1999, pp. 1–18.

Bernard, Henri, "Alcoolisme et antialcoolisme en France au XIXe siècle: Autour de Magnus Huss", *Histoire, Economie & Société*, 3 (1984), pp. 609–28.

Berrios, German and Hugh Freeman, "Introduction", in German Berrios and Hugh Freeman (eds), *150 Years of British Psychiatry, 1841–1991*, London: Gaskell, 1991, pp. ix–xiv.

Berrios, German and Roy Porter, "Epilepsy", in German Berrios and Roy Porter (eds), *A History of Clinical Psychiatry: The Origin and History of Psychiatric Disorders*, London: Athlone, 1995, pp. 147–73.

Berrios, German, "Memory and the cognitive paradigm", in Robin Murray and Trevor Turner (eds), *Lectures on the History of Psychiatry: The Squibb Series*, London: Gaskell, 1990, pp. 194–211.

Berrios, German, Roy Porter and Tom Kitwood, "Dementia", in German Berrios and Roy Porter (eds), *A History of Clinical Psychiatry: The Origin and History of Psychiatric Disorders*, London: Athlone, 1995, pp. 34–71.

Berrios, German, *The History of Mental Symptoms: Descriptive Psychopathology since the Nineteenth Century*, Cambridge: Cambridge University Press, 1995.

Beveridge, Allan, "Life in the asylum: Patients' letters from Morningside, 1873-1908", *History of Psychiatry*, 9 (1998), pp. 431–68.

Beveridge, Allan, "Madness in Victorian Edinburgh: A study of patients admitted to the Royal Edinburgh Asylum under Thomas Clouston 1873-1908", part 2, *History of Psychiatry*, 6 (1995), pp. 431–51.

Blue, Amy, *Η Δημιουργία της Ελληνικής Ψυχιατρικής [The Creation of Greek Psychiatry]*, Athens: Εξάντας [Exandas], 1999.

Blustein, Bonnie Ellen, "'A hollow square of psychological science': American neurologists and psychiatrists in conflict", in Andrew Scull (ed.), *Madhouses, Mad-Doctors, and Madmen. The Social History of Psychiatry in the Victorian Era*, London: Athlone Press, 1981, pp. 241–70.

Bonaïti, Muriel, "Histoire de l'asile XIXe siècle: D'un lieu unique de soin aux premières alternatives à l'hospitalisation", PhD thesis, Université de Lyon, 2007.

Bourke, Joanna, *Working-Class Cultures in Britain 1890-1960: Gender, Class and Ethnicity*, London: Routledge, 1994.

Bournova, Evgenia, *Οι Κάτοικοι των Αθηνών (1900–1960) – Δημογραφία [The Residents of Athens (1900–1960) – Demography]*, Athens: Εθνικό και Καποδιστριακό Πανεπιστήμιο Αθηνών-Τμήμα Οικονομικών Επιστημών [National and Kapodistrian University of Athens-Department of Economics], 2016.

Boutzovi, Aleka, "Γυναικείο κίνημα. Όψεις και δράσεις, 1909–1922" ["Women's movement. Aspects and actions, 1909–1922"], in Vasilis Panayotopoulos (ed.), *Ιστορία του Νέου Ελληνισμού, 1770–2000 [History of Modern Hellenism, 1770–2000]*, v. 6, Athens: Ελληνικά Γράμματα [Ellinika Grammata], 2003, pp. 283–92.

Budde, Gunilla-Friederike, *Auf dem Weg ins Bürgerleben: Kindheit und Erziehung in Deutschen und Englischen Bürgerfamilien, 1840-1914*, Gottingen: Vandenhoeck & Ruprecht, 1994.

Busfield, Joan, "Class and gender in twentieth-century British psychiatry: Shell-shock and psychopathic disorder", in Jonathan Andrews and Anne Digby (eds), *Sex and Seclusion, Class and Custody: Perspectives on Gender and Class in the History of British and Irish Psychiatry*, Amsterdam: Rodopi, 2004, pp. 295–322.

Busfield, Joan, "The female malady? Men, women and madness in nineteenth-century Britain", *Sociology*, 28 (1994), pp. 259–77.

Busfield, Joan, *Men, Women, and Madness: Understanding Gender and Mental Disorder*, Basingstoke: Macmillan, 1996.

Bynum, William Frederick, Roy Porter and Michael Shepherd, "Introduction", in William Frederick Bynum, Roy Porter and Michael Shepherd (eds), *The Anatomy of Madness: Essays in the History of Psychiatry,* vol. 2, London: Tavistock, 1985, pp. 1–16.

Bynum, William Frederick, Roy Porter and Michael Shepherd, *The Anatomy of Madness: Essays in the History of Psychiatry,* 3 vols, London: Tavistock, 1985–88.

Bynum, William, "Alcoholism and degeneration in 19th-century European medicine and psychiatry", *British Journal of Addiction*, 79 (1984), pp. 59–70.

Bynum, William, "The nervous patient in 18th- and 19th-century Britain: The psychiatric origins of British neurology", in Robin Murray and Trevor Turner (eds), *Lectures on the History of Psychiatry: The Squibb Series*, London: Gaskell, 1990, pp. 115–27.

Campbell, Brad, "The making of 'American': Race and nation in neurasthenic discourse", *History of Psychiatry*, 2007 (18), pp. 157–78.

Castel, Robert, *The Regulation of Madness: The Origins of Incarceration in France*, Cambridge: Polity, 1988.

Charitable Society of Athens, *Γηροκομείο 1864–1964 [Old People's Home 1864–1964]*, s.l.: s.n., s.d.

Chesler, Phyllis, *Women and Madness*, Garden City, N.Y.: Doubleday, 1972.

Chesters, Janice, "A horror of the asylum or of the home: Women's stories 1880-1910", in Catherine Coleborne and Dolly MacKinnon (eds), *Madness in Australia: Histories, Heritage and the Asylum*, St. Lucia: University of Queensland Press, 2003, pp. 135–44.

Clogg, Richard, *A Concise History of Greece*, 4th edition, Cambridge: Cambridge University Press, 2021.

Coleborne, Catharine, "'His brain was wrong, his mind astray': Families and the language of insanity in New South Wales, Queensland, and New Zealand, 1880s-1910", *Journal of Family History*, 31 (2006), pp. 45–65.

Coleborne, Catharine, *Madness in the Family: Insanity and Institutions in the Australasian Colonial World 1860-1914*, New York: Palgrave Macmillan, 2010.

Coleborne, Catharine, *Reading "Madness": Gender and Difference in the Colonial Asylum in Victoria, Australia, 1848–1888*, Perth: API Network, 2007.

Coleborne, Catherine, "'She does up her hair fantastically': The production of femininity in patient case books of the lunatic asylum in 1860s Victoria", in Jane Long (ed.), *Forging Identities: Bodies, Gender and Feminist History*, Nedlands: University of Western Australia Press, 1997, pp. 47–68.

Crowther, Margaret Anne, "Poverty, health and welfare", in William Hamish Fraser and Robert John Morris (eds), *People and Society in Scotland, 1830–1914*, v. 2, Edinburgh: J. Donald, 1990, pp. 265–89.

Dal Zilio, Samuel, "'Réclamée par sa famille contre l'avis des médecins': Pouvoir et implication des familles de patients dans les processus d'internement psychiatrique à l'asile de Schaerbeek entre 1850 et 1914", *Journal of Belgian History*, 51 (2021), pp. 100–24.

Davidoff, Leonore and Catherine Hall, *Family Fortunes: Men and Women of the English Middle Class, 1780-1850*, London: Hutchinson, 1987.

Davis, Gayle, *"The Cruel Madness of Love": Sex, Syphilis and Psychiatry in Scotland, 1880-1930*, Amsterdam: Rodopi, 2008.

Delvaux, Martine, *Femmes Psychiatrisées, Femmes Rebelles: De l'Étude de Cas à la Narration Autobiographique*, Paris: Le Plessis-Robinson, Institut Synthelabo, 1998.

Deniker, Pierre and Jean-Pierre Olie, *Fou, Moi? La Psychiatrie Hier et Aujourd'hui*, Paris: O. Jacob, 1998.

Deutsch, Albert, *The Mentally Ill in America: A History of their Care and Treatment from Colonial Times*, Garden City, N.Y.: Doubleday, Doran & Company, 1937.

Devine, Thomas Martin, *The Scottish Nation 1700-2000*, London: Penguin Books, 1999.

Digby, Anne, "Quantitative and qualitative perspectives on the asylum", in Roy Porter and Andrew Wear (eds), *Problems and Methods in the History of Medicine*, London: Croom Helm, 1987, pp. 153–74.

Doody, G. A., Allan Beveridge and E. C. Johnstone, "Poor and mad: A study of patients admitted to the Fife and Kinross District Asylum between 1874 and 1899", *Psychological Medicine*, 26 (1996), pp. 887–98.

Dörner, Klaus, *Bürger und Irre: Zur Sozialgeschichte und Wissenschaftssoziologie der Psychiatrie*, Frankfurt am Main: Fischer Taschenbuch Verlag, 1975.

Dowbiggin, Ian, "Back to the future: Valentin Magnan, French psychiatry, and the classification of mental diseases, 1885-1925", *Social History of Medicine*, 9 (1996), pp. 383–408.

Dowbiggin, Ian, "Delusional diagnosis? The history of paranoia as a disease concept in the modern era", History of Psychiatry, 11 (2000), pp. 37–69.

Dwyer, Ellen, *Homes for the Mad: Life Inside Two Nineteenth-Century Asylums*, New Brunswick, N.J.: Rutgers University Press, 1987.

Eghigian, Greg (ed.), *The Routledge History of Madness and Mental Health*, New York: Routledge, 2017.

Ellenberger, Henri, "La conférence de Freud sur l' hysterie masculine" [1968], in Élizabeth Roudinesco (ed.), *Médecines de l' Âme: Essais d' Histoire de la Folie et des Guérisons Psychiques*, Paris: Fayard, 1995, pp. 207–25.

Ellenberger, Henri, *The Discovery of the Unconscious: The History and Evolution of Dynamic Psychiatry*, London: Fontana Press, 1994.

Ellis, Robert, "The asylum, the Poor Law, and a reassessment of the four-shilling grant: Admissions to the county asylums of Yorkshire in the nineteenth century", *Social History of Medicine*, 19 (2006), pp. 55–71.

Engstrom, Eric, *Clinical Psychiatry in Imperial Germany: A History of Psychiatric Practice*, Ithaca: Cornell University Press, 2003.

Erickson, Edward, "Punishing the mad bomber: Questions of moral responsibility in the trials of French anarchist terrorists, 1886-1897", *French History*, 22 (2008), pp. 51–73.

Fauvel, Aude, "Le crime de Clermont et la remise en cause des asiles en 1880", *Revue d'Histoire Moderne et Contemporaine*, 49 (2002), pp. 195–216.

Fauvel, Aude, "Madness: A 'female malady'? Women and psychiatric institutionalisation in France", in Patrice Bourdelais and John Chircop (eds), *Vulnerabilities, Social Inequalities and Health in Perspective*, Évora: Ediçoes Colibri, 2010, pp. 61–75.

Fauvel, Aude, "Psychiatrie", *Bulletin de la Société Française pour l'Histoire des Sciences de l'Homme*, 29 (2006), pp. 43–51.

Feeley, Malcolm and Deborah Little, "The vanishing female: The decline of women in the criminal process, 1687-1912", in Robert Weiss (ed.), *The Social History of Crime, Policing and Punishment*, Aldershot: Ashgate, 1999, pp. 267–305.

Filandrianos, Kostas, *Δημόσιο Ψυχιατρείο Αθηνών. Το Δαφνί, μια Φανταστική Πολιτεία [The Public Psychiatric Hospital of Athens. Dafni, an Imaginary City]*, Athens: s.n., 1977.

Finnane, Mark, "Asylums, families and the state", *History Workshop Journal*, 20 (1985), pp. 134–48.

Finnane, Mark, "From dangerous lunatic to human rights? The law and mental illness in Australian history", in Catherine Coleborne and Dolly MacKinnon (eds), *Madness in Australia: Histories, Heritage and the Asylum*, St. Lucia: University of Queensland Press, 2003, pp. 23–36.

Finnane, Mark, *Insanity and the Insane in Post-Famine Ireland*, London: Croom Helm, 1981.

Fish, Frank, "David Skae, M.D., F.R.C.S, founder of the Edinburgh school of psychiatry", *Medical History*, 9 (1965), pp. 36–53.

Forrester, John, "If p, then what? Thinking in cases", *History of the Human Sciences*, 9 (1996), pp. 1–25.

Forth, Christopher, "Neurasthenia and manhood in fin-de-siècle France", in Marijke Gijswijt-Hofstra and Roy Porter (eds), *Cultures of Neurasthenia from Beard to the First World War*, Amsterdam, Rodopi, 2001, pp. 329–61.

Foucault, Michel, *Le Pouvoir Psychiatrique: Cours au Collège de France (1973–1974)*, ed. by J. Lagrange, Paris: Seuil/Gallimard, 2003.

Foucault, Michel, *Les anormaux: Cours au Collège de France Année 1974-1975*, Paris: Seuil/Gallimard, 1999.

Foucault, Michel, *Madness and Civilization. A History of Insanity in the Age of Reason*, transl. Richard Howard, New York: Vintage, 1973.

Foucault, Michel, *Surveiller et Punir: Naissance de la Prison*, Paris: Gallimard, 1975.

Foucault, Michel, *The Birth of the Clinic: An Archaeology of Medical Perception*, transl. A. M. Sheridan, London: Tavistock, 1976.

Fountanopoulos, Kostas, "Η μισθωτή εργασία" ["Paid work"], in Christos Hadziiossif (ed.), *Ιστορία της Ελλάδας του 20ού αιώνα [History of Greece in the 20th Century]*, v. A1, Athens: Βιβλιόραμα [Vivliorama], 1998, pp. 87–121.

Fournaraki, Eleni, "Περί μορφώσεως 'χρηστών μητέρων' και εκπαιδεύσεως 'μελλόντων πολιτών': Έμφυλοι λόγοι στην ελληνική εκπαίδευση τον 19ο αιώνα" ["On the formation of 'righteous mothers' and the education of 'future citizens': Gendered discourses in Greek education in the 19th century"], in Vasiliki Theodorou and Vasiliki Kontoyianni (eds), *Το Παιδί στη Νεοελληνική Κοινωνία, 19ος-20ός αιώνας. Αξίες, Αναπαραστάσεις, Αποτυπώσεις [The Child in Modern Greek Society, 19th-20th century. Values, Representations, Impressions]*, Athens: Ελληνικό Λογοτεχνικό και Ιστορικό Αρχείο [Hellenic Literary and Historical Archive], 1999, pp. 73–120.

Fournaraki, Eleni, *Εκπαίδευση και Αγωγή των Κοριτσιών: Ελληνικοί Προβληματισμοί (1830–1910): Ένα Ανθολόγιο [Schooling and Education of Girls: Greek Reflections (1830–1910): An Anthology]*, Athens: Γενική Γραμματεία Νέας Γενιάς [General Secretariat for Youth], 1987.

Gamwell, Lynn and Nancy Tomes, *Madness in America: Cultural and Medical Perceptions of Mental Illness before 1914*, Binghamton: Cornell University Press, 1995.

Garton, Stephen, "Asylum histories: Reconsidering Australia's lunatic past", in Catherine Coleborne and Dolly MacKinnon (eds), *Madness in Australia: Histories, Heritage and the Asylum*, St. Lucia: University of Queensland Press, 2003, pp. 11–22.

Gateaux-Mennecier, Jaqueline, *Bourneville et l'Enfance Aliénée*, Paris: Centurion, 1989.

Gelfand, Toby, "Neurologist or psychiatrist? The public and private domains of Jean-Martin Charcot", *Journal of the History of the Behavioral Sciences*, 36 (2000), pp. 215–29.

Gijswijt-Hofstra, Marijke and Roy Porter (eds), *Cultures of Neurasthenia from Beard to the First World War*, Amsterdam: Rodopi, 2001.

Gijswijt-Hofstra, Marijke, "Introduction", in Marijke Gijswijt-Hofstra and Roy Porter (eds), *Cultures of Neurasthenia from Beard to the First World War*, Amsterdam: Rodopi, 2001, pp. 1–30.

Gilman, Sander, Helen King, Roy Porter, G. S. Rousseau, Elaine Showalter (eds), *Hysteria beyond Freud*, Berkeley: University of California Press, 1993.

Goffman, Erving, *Asylums: Essays on the Social Situation of Mental Patients and other Inmates,* Garden City, N.Y.: Anchor Books, 1961.

Goldstein, Jan, "The hysteria diagnosis and the politics of anticlericalism in late nineteenth-century France", *The Journal of Modern History*, 54 (1982), pp. 209–39.

Goldstein, Jan, *Console and Classify: The French Psychiatric Profession in the Nineteenth Century*, Cambridge: Cambridge University Press, 1987.

Goodman, Helen, "'Madness and masculinity'. Male patients in London asylums and Victorian culture", in Thomas Knowles and Serena Trowbridge, *Insanity and the Lunatic Asylum in the Nineteenth Century*, London: Pickering & Chatto, 2015, pp.149–66.

Gotsinas, Kostis, "'Disciples of Asclepius' or 'advocates of Hermes'? Psychiatrists and alcohol in early-twentieth-century Greece", in Waltraud Ernst and Thomas Müller (eds), *Alcohol, Psychiatry and Society*, Manchester: Manchester University Press, 2022, pp. 220–47.

Gotsinas, Kostis, *Ένα Τελευταίο Ποτήρι; Αλκοόλ και Αντιαλκοολισμός στην Ελλάδα (1870–1940) [One Last Cup? Alcohol and Anti-Alcoholism in Greece (1870–1940)]*, Research Centre for the Humanities Digital Library, 2020.

Gotsinas, Kostis, *Επί της Ουσίας. Ιστορία των Ναρκωτικών στην Ελλάδα (1875–1950) [In Substance. History of Drugs in Greece (1875–1950)]*, Heraclio: Πανεπιστημιακές Εκδόσεις Κρήτης [Crete University Press], 2021.

Granshaw, Lindsay and Roy Porter (eds), *The Hospital in History*, London: Routledge, 1989.

Granshaw, Lindsay, "The rise of the modern hospital in Britain", in Andrew Wear (ed.), *Medicine in Society: Historical Essays*, Cambridge: Cambridge University Press, 1992. pp. 197–218.

Grant, Julia, "Parent-child relations in Western Europe and North America, 1500-present", in Paula S. Fass (ed.), *The Routledge History of Childhood in the Western World*, Routledge, London, 2013, pp. 103–24.

Grob, Gerald, "The history of the asylum revisited: Personal reflections", in Mark Micale and Roy Porter (eds), *Discovering The History of Psychiatry*, New York: Oxford University Press, 1994, pp. 260–81.

Grob, Gerald, *Mental Institutions in America: Social Policy to 1875*, New York: Free Press, 1973.

Guenther, Katja, *Localization and Its Discontents: A Genealogy of Psychoanalysis and the Neuro Disciplines,* Chicago: University of Chicago Press, 2015.

Hadziiossif, Christos, "Εισαγωγή" ["Introduction"], in Christos Hadziiossif (ed.), *Ιστορία της Ελλάδας του 20ού Αιώνα [History of Greece in the 20th Century]*, v. A1, Athens: Βιβλιόραμα [Vivliorama], 1998, pp. 9–39.

Hadziiossif, Christos, "Η μπελ επόκ του κεφαλαίου" ["The belle époque of capital"], in Christos Hadziiossif (ed.), *Ιστορία της Ελλάδας του 20ού Αιώνα [History of Greece in the 20th Century]*, v. A1, Athens: Βιβλιόραμα [Vivliorama], 1998, pp. 309–49.

Hall, Lesley, "Interpreting abnormal psychology in the late nineteenth century: William James's spiritual crisis", in Waltraud Ernst (ed.), *Histories of the Normal and the Abnormal: Social and Cultural Histories of Norms and Normativity*, London: Routledge, 2006.

Haller, John and Robin Haller, *The Physician and Sexuality in Victorian America*, Urbana: University of Illinois Press, 1974.

Hantzaroula, Pothiti, *Σμιλεύοντας την Υποταγή. Οι Έμμισθες Οικιακές Εργάτριες στην Ελλάδα το Πρώτο Μισό του Εικοστού Αιώνα [Forging Submission. Paid Domestic Workers in Greece in the First Half of the Twentieth Century]*, Athens: Παπαζήσης [Papazisis], 2012.

Harlaftis, Gelina, "Ο εμπορικός στόλος στους πέντε ωκεανούς και τις πέντε ηπείρους" ["The merchant fleet in the five oceans and five continents"], in Vasilis Panayotopoulos (ed.), *Ιστορία του Νέου Ελληνισμού, 1770-2000 [History of Modern Hellenism, 1770-2000]*, v. 6, Athens: Ελληνικά Γράμματα [Ellinika Grammata], 2003, pp. 129–38.

Harrison, Brian, *Drink and the Victorians: The Temperance Question in England, 1815-1872*, London: Faber and Faber, 1971.

Harsin, Jill, "Gender, class, and madness in nineteenth-century France", *French Historical Studies*, 17 (1992), pp. 1048–70.

Harsin, Jill, "Syphilis, wives, and physicians: Medical ethics and the family in late nineteenth-century France",*French Historical Studies*, 16 (1989), pp. 72–95.

Heather Montgomery, *An Introduction to Childhood: Anthropological Perspectives on Children's Lives*, Chichester, West Sussex, U.K.: Wiley-Blackwell, 2009.

Hendrick, Harry, *Images of Youth: Age, Class and the Male Youth Problem, 1880-1920*, Oxford: Clarendon Press, 1990.

Hewitt, Jessie, *Institutionalizing Gender. Madness, the Family, and Psychiatric Power in Nineteenth-Century France*, Ithaca: Cornell University Press, 2020.

Higgs, Paul and Chris Gilleard, "Ageing, dementia and the social mind: Past, present and future perspectives", *Sociology of Health & Illness*, 39: 2 (2017), pp. 175–81.

Hilton, Claire, "Age inclusive services or separate old age and working age services? A historical analysis from the formative years of old age psychiatry c.1940-1989", *Psychiatric Bulletin*, 39: 2 (2015), pp. 90–95.

Hirschmüller, Albrecht and Magda Whitrow, "The development of psychiatry and neurology in the nineteenth century", *History of Psychiatry*, 10 (1999), pp. 395–423.

Hodgkin, Katherine, *Madness in Seventeenth-Century Autobiography*, Basingstoke: Palgrave Macmillan, 2007.

Holden, Wendy, *Shell Shock*, London: Channel 4, 1998.

Houston, Robert Allan, *Madness and Society in Eighteenth-Century Scotland*, Oxford: Clarendon Press, 2000.

Hughes, John Starrett, "The madness of separate spheres: Insanity and masculinity in Victorian Alabama", in Mark Carnes and Clyde Griffen (eds), *Meanings for Manhood: Constructions of Masculinity in Victorian America*, Chicago: University of Chicago Press, 1990, pp. 53–66.

Hunter, Richard and Ida Macalpine, *Three Hundred Years of Psychiatry, 1535-1860: A History Presented in Selected English Texts*, Hartsdale, N.Y.: Carlisle Publications, 1982.

Ignatieff, Michael, "State, civil society and total institutions: A critique of recent social histories of punishment", in Stanley Cohen and Andrew Scull (eds), *Social Control and the State: Historical and Comparative Essays*, Oxford: Robertson, 1983, pp. 75–105.

Ion, R.M. and M.D. Beer, "The British reaction to dementia praecox 1893-1913", part 1, *History of Psychiatry*, 13 (2002), pp. 285–304.

Jackson, Mark, "'Grown-up children': Understandings of health and mental deficiency in Edwardian England", in Marijke Gijswijt-Hofstra and Hilary Marland (eds), *Cultures of Child Health in Britain and the Netherlands in the Twentieth Century*, Amsterdam and Atlanta: Rodopi, 2003, pp. 141–60.

Jacob, Françoise, "'Folie' et étude de cas. Jean Longuebeau, ouvrier cloutier, un délire à Toulouse, en 1836", in François-Olivier Touati (ed.), *Maladies, Médecines et Sociétés: Approches Historiques pour le Present: Actes du VIe Colloque d'Histoire au Present*, Paris: L' Harmattan, 1993, pp. 64–75.

James, Alisson, Chris Jenks and Alisson Prout, *Theorizing Childhood*, London: Polity, 2010.

Jenks, Chris, *Childhood*, London: Routledge, 2005.

Jones, Kathleen, *Lunacy, Law and Conscience 1744-1845*, London: Routledge Kegan Paul, 1955.

Jones, Kevin Matthew, *Psychological Classification and Diagnosis in Asylum Statistics, 1800-1948. The British Table of the Forms of Insanity*, Cham: Palgrave Macmillan, 2025.

Karadimou-Gerolympou, Aleka, "Πόλεις και πολεοδομία" ["Cities and urban planning"], in Christos Hadziiossif (ed.), *Ιστορία της Ελλάδας του 20ού Αιώνα [History of Greece in the 20th Century]*, v. A1, Athens: Βιβλιόραμα [Vivliorama], 1998, pp. 223–53.

Karamanolakis, Vangelis, "Η ανάδυση των ιδρυματικών θεσμών. Το Δρομοκαΐτειο φρενοκομείο (1880–1903)" ["The emergence of institutions. The Dromokaitio Lunatic Asylum (1880–1903)"], MA thesis, Athens, National and Kapodistrian University of Athens, 1997.

Karamanolakis, Vangelis, "Οι φρενοβλαβείς εν ταις οδοίς" ["The lunatics in the streets"], in *Η Πόλη στους Νεότερους Χρόνους: Μεσογειακές και Βαλκανικές*

Όψεις (19ος-20ός Αιώνας) [The City in Modern Times: Mediterranean and Balkan Aspects (19th-20th Century)], Proceedings of the 2nd International Conference of History, Athens: Εταιρεία Μελέτης Νέου Ελληνισμού [Society for the Study of Modern Hellenism], 2000, pp. 193–202.

Karamanolakis, Vangelis, "Το Δρομοκαΐτειο φρενοκομείο: 1887–1903. Όψεις της εγκατάστασης ενός ιδρυματικού θεσμού" ["The Dromokaitio Lunatic Asylum: 1887–1903. Aspects of the establishment of an institution"], *Μνήμων [Mninon]*, 20 (1998), pp. 45–66.

Karavatos, Athanasios, "Η εισαγωγή στην Ελλάδα του όρου 'πρωτόγονος (ή πρώιμος) άνοια'. Επιδράσεις της γαλλικής και γερμανικής ψυχιατρικής σχολής" ["The introduction in Greece of the term 'primitive (or early) dementia'. Influences of the French and German psychiatric school"], *Ψυχιατρική*, [*Psychiatry*] 6 (1995), pp. 136–44.

Karavatos, Athanasios, Dimitris Ploumpidis and Yorgos Christodoulou, "Οικονομάκης Μιλτιάδης" ["Economakis Miltiadis"], in Athanasios Karavatos, Dimitris Ploumpidis and Yorgos Christodoulou (eds), *Ανθολόγιο Ελληνικών Ψυχιατρικών Κειμένων [Anthology of Greek Psychiatric Texts]*, Athens: Ελληνική Ψυχιατρική Εταιρεία [Hellenic Psychiatric Association], 2006, pp. 593–94.

Karavatos, Thanasis, "Η γένεση της ψυχιατρικής και του ασύλου. Οι μυθοποιήσεις της πραγματικότητας και ο πραγματικός μύθος" ["The genesis of psychiatry and the asylum. The mythologisations of reality and the real myth"], *Ο Πολίτης [The Citizen]*, 78 (1987), 52–55.

Karavatos, Thanasis, *Τα Πρώτα Βήματα της Ψυχολογίας στην Ελλάδα. Η «Ψυχολογία» του Γεωργίου Σερούϊου [The First Steps of Psychology in Greece. The "Psychology" of Yeorgios Serouios]*, Athens: Οδυσσέας [Odysseas], 1991.

Kardamitsi-Adami, Maro, "Ο αστικός χώρος, 1871–1909. Πολεοδομία-αρχιτεκτονική" ["The urban space, 1871–1909. Urban planning-architecture"], in Vasilis Panayotopoulos (ed.), *Ιστορία του Νέου Ελληνισμού, 1770–2000 [History of Modern Hellenism, 1770–2000]*, v. 5, Athens: Ελληνικά Γράμματα [Ellinika Grammata], 2003, pp. 107–26.

Karydaki, Danae (ed.), *Η Λέρος στο Επίκεντρο και το Περιθώριο [Leros in the Spotlight and the Margin]*, Athens: Ψηφίδες [Psiphides], 2020.

Kaufmann, Doris, "Neurasthenia in Wilhelmine Germany: Culture, sexuality and the demands of nature", in Marijke Gijswijt-Hofstra and Roy Porter (eds), *Cultures of Neurasthenia from Beard to the First World War*, Amsterdam: Rodopi, 2001, pp. 161–76.

Kazolea-Tavoulari, Panayota, "Άσυλο φρενοβλαβών Σύρου: Ιστορική διερεύνηση προνοιακών θεσμών και κοινωνικές αναπαραστάσεις περί ψυχικής νόσου" ["Syros Lunatic Asylum: Historical investigation of welfare institutions and social representations of mental illness"], *Τα Ιστορικά [Ta Historica]*, 19 (2002), pp. 377–98.

Keller, Richard, *Colonial Madness: Psychiatry in French North Africa*, Chicago: University of Chicago Press, 2007.

Kevles, Daniel, *In the Name of Eugenics: Genetics and the Uses of Human Heredity*, New York: Knopf, 1985.

Killen, Andreas, *Berlin Electropolis: Shock, Nerves, and German Modernity*, Berkeley: University of California Press, 2006.

Kim, H. Yumi, *Madness in the Family: Women, Care, and Illness in Japan*, Oxford, New York: Oxford University Press, 2022.

Kitromilidis, Paschalis, *Eleftherios Venizelos: The Trials of Statesmanship*, Edinburgh: Edinburgh University Press, 2006.

Kokkinos, Yeoryios, "Η αμφιθυμία για την πόλη. Η ελληνική διανόηση στην καμπή του 19ου αιώνα και το αστικό φαινόμενο" ["Ambivalence about the city. Greek intellectuals at the turn of the 19th century and the urban phenomenon"], in*Η Πόλη στους Νεότερους Χρόνους: Μεσογειακές και Βαλκανικές Όψεις (19ος-20ός Αιώνας) [The City in Modern Times: Mediterranean and Balkan Aspects (19th-20th Century))]*, Proceedings of the 2nd International Conference of History, Athens: Εταιρεία Μελέτης Νέου Ελληνισμού [Society for the Study of Modern Hellenism], 2000, pp. 597–614.

Kokkinos, Yoryos, *"Άξια" και "Ανάξια" Ζωή. Ευγονική, Εκφυλισμός, Βιοπολιτική: Ο Γιατρός στο Ρόλο του Κοινωνικού Θεραπευτή και του Εθνικού Αναμορφωτή ["Worthy" and "Unworthy" Life. Eugenics, Degeneration, Biopolitics: The Doctor in the Role of Social Healer and National Reformer]*, Athens: Ταξιδευτής [Taxideftis], 2021.

Korasidou, Maria, *Οι Άθλιοι των Αθηνών και οι Θεραπευτές τους: Φτώχεια και Φιλανθρωπία στην Ελληνική Πρωτεύουσα τον 19ο Αιώνα [The Miserable of Athens and their Healers: Poverty and Charity in the Greek Capital in the 19th Century]*, Athens: Κέντρο Νεοελληνικών Ερευνών, Εθνικό Ίδρυμα Ερευνών [Centre for Modern Greek Studies, National Hellenic Research Foundation], 1995.

Korasidou, Maria, *Όταν η Αρρώστια Απειλεί: Επιτήρηση και Έλεγχος της Υγείας του Πληθυσμού στην Ελλάδα του 19ου Αιώνα [When the Disease Threatens: Surveillance and Control of Population Health in 19th-Century Greece]*, Athens: Τυπωθήτω [Tipothito], 2002.

Kritsotaki, Despo, "Changing psychiatry or changing society? The "Motion for the Rights of the 'Mentally Ill' in Greece, 1980–1990", *Journal for the History of Medicine and Allied Sciences*, 2021.

Kritsotaki, Despo, "Σεξουαλικότητα και επιστήμες 'ψ' στην Ελλάδα: Έφηβες και νέες γυναίκες σε μια υπηρεσία ψυχικής υγείας, 1960-1980" ["Sexuality and the 'psy' sciences in Greece: Adolescent girls and young women in a mental health service, 1960-1980"], in Dimitra Vassiliadou and Glafki Gkotsi (eds), *Ιστορίες για τη Σεξουαλικότητα [Histories about Sexuality]*, Athens: Θεμέλιο [Themelio], 2020, pp. 81–100.

Kritsotaki, Despo, *"Ψυχική Υγιεινή", Κοινωνική Πρόνοια και Ψυχιατρική Μεταρρύθμιση στη Μεταπολεμική Ελλάδα. Το Κέντρο Ψυχικής Υγιεινής και Ερευνών, 1956–1978 ["Mental Hygiene", Social Welfare and Psychiatric Reform in Post-War Greece. The Centre for Mental Health and Research, 1956–1978]*, Athens: Πεδίο [Pedio], 2016.

Laing, Ronald, *The Politics of Experience and the Bird of Paradise*, Harmondsworth: Penguin, 1967.

Lanteri-Laura, Georges, "La chronicité dans la psychiatrie moderne française. Note d' histoire histoire théorique et sociale", *Annales. Économies, Sociétés, Civilisations*, 27 (1972), pp. 548–68.

Lapeyrière, Solange, "Celles qui sortent et celles qui restent. 'Carrières asilaires' des femmes internées dans les asiles en France au XIX siècle", *Clio. Femmes, Genre, Histoire*, 51 (2020), pp. 283–308.

Laqueur, Thomas, *Making Sex: Body and Gender from the Greeks to Freud*, Cambridge, Mass.: Harvard University Press, 1992.

Lawton, Julia, "Lay experiences of health and illness: Past research and future agendas", *Sociology of Health & Illness,* 25 (2003), pp. 23–40.

Le Bras, Anatole, "Troubles dans la masculinité", *Histoire, Médecine et Santé*, 23 (2023), pp. 55-69.

Le Bras, Anatole, *Aliénés - Une Histoire Sociale de la Folie au XIXe Siècle*, Paris: CNRS Éditions, 2024.

Le Bras, Anatole, *Un Enfant à l'Asile - Vie de Paul Taesch (1874-1914)*, Paris: CNRS Éditions, 2018.

Le Gaufey, Guy and Gérard Bleandonu, "Naissance des asiles d' aliénés (Auxerre-Paris)", *Annales*, 1975 (30), pp. 93–121.

Leontidou, Lila, *Πόλεις της Σιωπής. Εργατικός Εποικισμός της Αθήνας και του Πειραιά, 1909–1940 [Cities of Silence. Labour Settlement of Athens and Piraeus, 1909–1940]*, Athens: Πολιτιστικό Ίδρυμα Ομίλου Πειραιώς [Piraeus Bank Group Cultural Foundation], 2013.

Lerner, Paul, *Hysterical Men: War, Psychiatry, and the Politics of Trauma in Germany, 1890-1930*, Ithaca, N.Y.: Cornell University Press, 2003.

Levine-Clark, Marjorie, "'Embarrassed circumstances': Gender, poverty, and insanity in the West Riding of England in the early Victorian years", in Jonathan Andrews and Anne Digby (eds), *Sex and Seclusion, Class and Custody: Perspectives on Gender and Class in the History of British and Irish Psychiatry*, Amsterdam: Rodopi, 2004, pp. 123–48.

Levine-Clark, Marjorie, "Dysfunctional domesticity: Female insanity and family relationships among the West Riding poor in the mid-nineteenth century", *Journal of Family History*, 25 (2000), pp. 341–61.

Liakos, Antonis, *Εργασία και Πολιτική στην Ελλάδα του Μεσοπολέμου: Το Διεθνές Γραφείο Εργασίας και η Ανάδυση των Κοινωνικών Θεσμών [Labour and Politics in interwar Greece: The International Labour Office and the Emergence of*

Social Institutions], Athens: Εμπορική Τράπεζα της Ελλάδος [Commercial Bank of Greece], 1993.

Lixouriotis, Yannis, *Κοινωνικές και Νομικές Αντιλήψεις για το Παιδί τον Πρώτο Αιώνα του Νεοελληνικού Κράτους [Social and Legal Perceptions of the Child in the First Century of the Modern Greek State]*, Athens/Ioannina: Δωδώνη [Dodoni], 1986.

Loughran, Tracey, "Hysteria and neurasthenia in pre-1914 British medical discourse and in histories of shell-shock", *History of Psychiatry*, 19 (2008), pp. 25–46.

Loukos, Christos, "Η ελληνική πρωτεύουσα, 1890–1912" ["The Greek capital, 1890–1912"], in Christina Koulouri (ed.), *Αθήνα, Πόλη των Ολυμπιακών Αγώνων 1896–1906 [Athens, City of the Olympic Games 1896–1906]*, Athens: Διεθνής Ολυμπιακή Ακαδημία [International Olympic Academy], 2004, pp. 55–98.

Loukos, Christos, *Πεθαίνοντας στη Σύρο τον 19ο αιώνα. Οι Μαρτυρίες των Διαθηκών [Dying in Syros in the 19th century. The Testimonies of the Testaments]*, Heraklion: Πανεπιστημιακές Εκδόσεις Κρήτης [Crete University Press], 2000.

Luckins, Tanja, "'Crazed with grief?' The asylum and the Great War in Australia", in Catherine Coleborne and Dolly MacKinnon (eds), *Madness in Australia: Histories, Heritage and the Asylum*, St. Lucia: University of Queensland Press, 2003, pp. 169–80.

Lunbeck, Elizabeth, *The Psychiatric Persuasion: Knowledge, Gender, and Power in Modern America*, Princeton, N.J.: Princeton University Press, 1994.

Lutz, Tom, "Varieties of medical experience: Doctors and patients, psyche and soma in America", in Marijke Gijswijt-Hofstra and Roy Porter (eds), *Cultures of Neurasthenia from Beard to the First World War*, Amsterdam: Rodopi, 2001, pp 51–76.

MacDonald, Michael, "Madness, suicide, and the computer", in Roy Porter and Andrew Wear (eds), *Problems and Methods in the History of Medicine*, London: Croom Helm, 1987, pp. 207–29.

MacDonald, Michael, *Mystical Bedlam: Madness, Anxiety, and Healing in Seventeenth-Century England*, Cambridge: Cambridge University Press, 1981.

MacKenzie, Charlotte, "Social factors in the admission, discharge, and continuing stay of patients at Ticehurst Asylum, 1845–1917", in William Frederick Bynum, Roy Porter and Michael Shepherd (eds), *The Anatomy of Madness: Essays in the History of Psychiatry,* v. 2, London: Tavistock, 1985, pp. 147–74.

Madden, J. S., "Substance use disorders: Clinical section", in German Berrios and Roy Porter (eds), *A History of Clinical Psychiatry: The Origin and History of Psychiatric Disorders*, London: Athlone, 1995, pp. 656–67.

Majerus, Benoît, "Making sense of the 'chemical revolution'. Patients' voices on the introduction of neuroleptics in the 1950s, *Medical History* 60 (2016), pp. 54–66.

Majerus, Benoît, *Parmi les Fous: Une Histoire Sociale de la Psychiatrie au XXe Siècle*, Rennes: Presses Universitaires de Rennes, 2019.

Makras, Kostas, "'The poison that upsets my reason': Men, madness and drunkenness in the Victorian period", in Thomas Knowles and Serena Trowbridge, *Insanity and the Lunatic Asylum in the Nineteenth Century*, London: Pickering & Chatto, 2015, pp. 135–48.

Makrinioti, Dimitra, *Η Παιδική Ηλικία στα Αναγνωστικά Βιβλία, 1834–1919 [Childhood in Literacy Books, 1834–1919]*, Athens/Ioannina: Δωδώνη [Dodoni], 1986.

Mansker, Andrea, *Sex, Honor and Citizenship in Third Republic France*, London: Palgrave Macmillan, 2011.

Margadant, Jo Burr, "Introduction: Constructing selves in historical", in Jo Burr Margadant (ed.), *The New Biography: Performing Femininity in Nineteenth-Century France*, Berkeley: University of California Press, 2000, 1–32.

Marland, Hilary, "Disappointment and desolation: Women, doctors and interpretations of puerperal insanity in the nineteenth century", *History of Psychiatry*, 14 (2003), pp. 303–20.

Marland, Hilary, *Dangerous Motherhood: Insanity and Childbirth in Victorian Britain*, Houndmills: Palgrave Macmillan, 2004.

Mastroyannis, Ioannis, *Ιστορία της Κοινωνικής Προνοίας της Νεωτέρας Ελλάδος, 1821–1960 [History of Social Welfare in Modern Greece, 1821–1960]*, Athens: s.n., 1960.

Matthews, Jill Julius, *Good and Mad Women: The Historical Construction of Femininity in Twentieth-Century Australia*, Sydney: George Allen & Unwin, 1984.

Mazower, Mark, "The messiah and the bourgeoisie: Venizelos and politics in Greece, 1909-1912", *The Historical Journal*, 35 (1992), pp. 885–904.

Mazumdar, Pauline, *Eugenics, Human Genetics, and Human Failings: The Eugenics Society, its Sources and its Critics in Britain*, London: Routledge, 1992.

McCandless, Peter, "Liberty and lunacy: The Victorians and wrongful confinement", *Journal of Social History*, 11 (1978), pp. 366–86.

McGovern, Constance, "The community, the hospital, and the working-class patient: The multiple uses of asylum in nineteenth century America", *Pennsylvania History*, 54 (1987), pp. 17–33.

Micale, Mark and Paul Lerner, "Trauma, psychiatry, and history: A conceptual and historiographical introduction", in Mark Micale and Paul Lerner (eds), *Traumatic Pasts: History, Psychiatry, and Trauma in the Modern Age, 1870-1930*, Cambridge: Cambridge University Press, 2001, pp. 1–30.

Micale, Mark,"Charcot and the idea of hysteria in the male: Gender, mental science, and medical diagnosis in late nineteenth-century France", *Medical History*, 34 (1990), pp. 363–411.

Micale, Mark, "The Salpêtrière in the age of Charcot: An institutional perspective on medical history in the late nineteenth century", *Journal of Contemporary History*, 20 (1985), pp. 703–31.

Micale, Mark, *Approaching Hysteria: Disease and its Interpretations*, Princeton, N.J.: Princeton University Press, 1995.

Micale, Mark, *Hysterical Men. The Hidden History of Male Nervous Illness*, Cambridge, MA: Harvard University Press, 2008.

Mills, James, "The mad and the past: Retrospective diagnosis, post-coloniality, discourse analysis and the asylum archive", *Journal of Medical Humanities*, 21 (2000), pp. 141–58.

Miron, Janet, *Prisons, Asylums, and the Public: Institutional Visiting in the Nineteenth Century*, Toronto: The University of Toronto Press, 2011.

Mooney, Graham and Jonathan Reinarz (eds), *Permeable Walls: Historical Perspectives on Hospital and Asylum Visiting*, Amsterdam: Rodopi, 2009.

Morrison, Hazel, "Constructing patient stories: 'Dynamic' case notes and clinical encounters at Glasgow's Gartnavel Mental Hospital, 1921–32", *Medical History* 60 (2016), pp. 67–86.

Moscucci, Ornella, *The Science of Woman: Gynaecology and Gender in England, 1800-1929*, Cambridge: Cambridge University Press, 1990.

Nenadic, Stana, "The Victorian middle classes", in William Hamish Fraser and Irene Maver (eds), *Glasgow, 1830–1912*, v. 2, Manchester: Manchester University Press, 1996, pp. 265–99.

Neurological Clinic, University of Athens, *Αιγινήτειον Νοσοκομείον 1905–2005 [Eginitio Hospital 1905–2005]*, Athens: s.n., 2005.

Norbert Elias, *The Civilizing Process: The History of Manners*, New York: Urizen Books, 1978 (first edition in 1939).

Oppenheim, Janet, *"Shattered Nerves": Doctors, Patients, and Depression in Victorian England*, New York: Oxford University Press, 1991.

Osborn, Matthew Warner, "Diseased imaginations: Constructing delirium tremens in Philadelphia, 1813-1832", *Social History of Medicine*, 19 (2006) pp. 191–208.

Ozarin, Lucy, "The American Neurological Association: Early days and relations with psychiatry", *Psychiatric News*, 42 (2007), p. 11.

Papastefanaki, Leda, "Ανάμεσα στην ιστορία της βιομηχανίας και την ιστορία της εργασίας. Η οπτική του φύλου στην ελληνική ιστοριογραφία" ["Between the history of industry and the history of labour. The gender perspective in Greek historiography"], in Glafki Gotsi, Androniki Dialeti and Eleni Fournaraki (eds), *Το Φύλο στην Ιστορία. Αποτιμήσεις και Παραδείγματα [Gender in History. Assessments and Examples]*, Athens: Ασίνη [Asini], 2015, pp. 79–106.

Papastefanaki, Leda, "Μισθωτή εργασία" ["Paid work"], in Kostas Kostis and Socrates Petmezas (eds), Η Ανάπτυξη της Ελληνικής Οικονομίας κατά τον 19ο αιώνα (1830-1914) [The Development of the Greek Economy in the 19th century (1830-1914)], Athens: Αλεξάνδρεια [Alexandria], 2006, pp. 253–91.

Pappis, Konstantinos, "Η διοίκηση του Δρομοκαΐτειου" ["The administration of the Dromokaitio"], in Nikolaos Tsikis (ed.), *Δρομοκαΐτειο Θεραπευτήριο, 1887–1987+15. Εκατό Δεκαπέντε Χρόνια Κοινωνικής Προσφοράς [Dromokaitio Sanatorium, 1887–1987+15. One Hundred Fifteen Years of Social Offer]*, Athens: s.n., 2001, pp. 223–42.

Pappis, Konstantinos, "Η διοικητική διεύθυνση του Δρομοκαΐτειου" ["The administrative directorate of the Dromokaitio"], in Nikolaos Tsikis (ed.), *Δρομοκαΐτειο Θεραπευτήριο, 1887–1987+15. Εκατό Δεκαπέντε Χρόνια Κοινωνικής Προσφοράς [Dromokaitio Sanatorium, 1887–1987+15. One Hundred Fifteen Years of Social Offer]*, Athens: s.n., 2001, pp. 243–50.

Parker R., A. Dutta, R. Barnes and T. Fleet, "County of Lancaster Asylum, Rainhill: 100 years ago and now", *History of Psychiatry*, 4 (1993), pp. 95–105.

Parssinen, Terry and Karen Kerner, "Development of the disease model of drug addiction in Britain, 1870-1926", *Medical History*, 24 (1980), pp. 275–96.

Perrot, Michelle, "Roles and characters", in Philippe Aries and Georges Duby (eds), *A History of Private Life*, v. 4, Cambridge, MA: Harvard University Press, 1987, pp. 167–259.

Petmezas, Socrates, "Αγροτική οικονομία" ["Rural economy"], in Christos Hadziiossif (ed.), *Ιστορία της Ελλάδας του 20ού Αιώνα [History of Greece in the 20th Century]*, v. A1, Athens: Βιβλιόραμα [Vivliorama], 1998, pp. 53–85.

Petmezas, Socrates, "Η δημογραφική συγκυρία: Η δεύτερη φάση της 'δημογραφικής μετάβασης' και η υπερατλαντική μετανάστευση" ["The demographic conjuncture: The second phase of the 'demographic transition' and transatlantic migration"], in Christos Hadziiossif (ed.), *Ιστορία της Ελλάδας του 20ού Αιώνα [History of Greece in the 20th Century]*, v. A1, Athens: Βιβλιόραμα [Vivliorama], 1998, pp. 41–51.

Pick, Daniel, *Faces of Degeneration: A European Disorder, c.1848-c.1918*, Cambridge: Cambridge University Press, 1989.

Pinell, Patrice, "Dégénérescence", in Dominique Lecourt et al. (eds), *Dictionnaire de la Pensée Médicale*, Paris: Presses Universitaires de France, 2004, p. 310–12.

Pizanias, Petros, *Οι Φτωχοί των Πόλεων: Η Τεχνογνωσία της Επιβίωσης στην Ελλάδα το Μεσοπόλεμο [The Urban Poor: The Expertise of Survival in Greece in the Interwar Period]*, Athens: Θεμέλιο [Themelio], 1993.

Ploumpidis, Dimitris, *Ιστορία της Ψυχιατρικής στην Ελλάδα. Θεσμοί, Ιδρύματα και Κοινωνικό Πλαίσιο 1850- 1920 [History of Psychiatry in Greece: Institutions, Hospitals and Social Framework]*, Athens: Εξάντας [Exandas], 1995.

Porter, Roy, "Madness and the family before Freud: The view of the mad-doctors", *Journal of Family History*, 23 (1998), pp. 159–72.

Porter, Roy, "The body and mind, the doctor and the patient: Negotiating hysteria", in Sander Gilman, Helen King, Roy Porter, G. S. Rousseau and Elaine Showalter (eds), *Hysteria Beyond Freud*, Berkeley: University of California Press, 1993, pp. 225–85.

Porter, Roy, "The drinking man's disease: The 'pre-history' of alcoholism in Georgian Britain", *British Journal of Addiction*, 80 (1985), pp. 385–96.

Porter, Roy, "The patient's view: Doing medical history from below", Theory and Society, 14 (1985), pp. 175–98.

Porter, Roy, *ASocial History of Madness: Stories of the Insane*, London: Weidenfeld and Nicolson, 1987.

Porter, Roy, *Madness. A Brief History*, Oxford: Oxford University Press, 2002.

Porter, Roy, *Mind Forg'd Manacles:A History of Madness in England from the Restoration to the Regency*, London: Athlone, 1987.

Postel, Jacques, *Éléments pour une Histoire de la Psychiatrie Occidentale*, Paris: L' Harmattan, 2007.

Postel, Jacques, *Genèse de la Psychiatrie: Les Premiers Écrits Psychiatriques de Philippe Pinel*, Paris: Le Sycomore, 1981.

Preka, Maria, "Αρσακειάδες δασκάλες (1880–1913): Ρητές και άρρητες οικογενειακές διαδρομές" ["Teachers of the Arsakeio School (1880–1913): Explicit and implicit family paths"], in Glafki Gotsi, Androniki Dialeti and Eleni Fournaraki (eds), *Το Φύλο στην Ιστορία. Αποτιμήσεις και Παραδείγματα [Gender in History. Assessments and Examples]*, Athens: Ασίνη [Asini], 2015, pp. 107–38.

Prestwich Patricia, "Family strategies and medical power: 'Voluntary' committal in a Parisian asylum, 1876–1914", in Roy Porter and David Wright (eds), *The Confinement of the Insane: International Perspectives, 1800–1965*, Cambridge: Cambridge University Press, 2003, pp. 79–99.

Prestwich, Patricia, "Drinkers, drunkards, and degenerates: The alcoholic population of a Parisian asylum, 1867-1914", *Histoire Sociale/Social History*, 27 (1994), pp. 321–35.

Prestwich, Patricia, "Family strategies and medical power: 'Voluntary' committal in a Parisian asylum, 1876-1914", *Journal of Social History*, 1994 (27), pp. 799–814.

Prestwich, Patricia, "Traiter les alcooliques à la Belle Époque: Le rêve du 'grand renfermement'", in Francois-Olivier Touati (ed.), *Maladies, Médecines et Sociétés: Approches Historiques pour le Present*. Actes du VIe Colloque, v. 2, Paris: L'Harmattan, 1993, pp. 218–23.

Prichard, James, *A Treatise on Insanity and Other Disorders Affecting the Mind*, London: Sherwood, Gilbert, and Piper, 1835.

Psarra, Angelika, "Μητέρα ή πολίτις: Ελληνικές εκδοχές της γυναικείας χειραφέτησης (1870-1920)" ["Mother or citizen: Greek versions of female emancipation (1870-1920)"], *Δίνη [Dini]* 9 (1997), pp. 239–57.

Quétel, Claude and Jean-Yves Simon, "L' aliénation alcoolique en France (XIXe siècle et 1ère moitié du XXe siècle)", *Histoire, Economie & Société*, 7 (1988), pp. 507–33.

Raptis, Kostas, "Αστικές τάξεις και αστικότητα στην Ευρώπη, 1789-1914: Προσανατολισμοί της σύγχρονης ιστοριογραφίας" ["Middle classes and middle-class culture in Europe, 1789-1914: Approaches in modern historiography"], *Μνήμων [Mnimon]*, 20 (1998), pp. 211–43.

Reed, Matt, "Making the case: The evolution of the case study in nineteenth-century French psychiatry", PhD thesis, Faculty of Claremont Graduate University, 2002.

Renvoize, Edward and Allan Beveridge, "Mental illness and the late Victorians: A study of patients admitted to three asylums in York, 1880-1884", *Psychological Medicine*, 19 (1989), pp. 19–28.

Renvoize, Edward, "The Association of Medical Officers of Asylums and the Hospitals for the Insane, the Medico-psychological Association and their presidents", in German Berrios and Hugh Freeman (eds), *150 Years of British Psychiatry, 1841-1991*, London: Gaskell, 1991, pp. 29–78.

Repousi, Maria, "Το φύλο των γυναικών. Έμφυλες ομάδες και γυναικείες διεκδικήσεις" ["The gender of women. Gendered groups and women's demands"], in Vasilis Panayotopoulos (ed.), *Ιστορία του Νέου Ελληνισμού, 1770–2000 [History of Modern Hellenism, 1770–2000]*, v. 5, Athens: Ελληνικά Γράμματα [Ellinika Grammata], 2003, pp. 187–96.

Ripa, Yannick, *Women and Madness: The Incarceration of Women in Nineteenth-Century France*, Cambridge: Polity Press, 1990.

Risse, Günter and John Harley Warner, "Reconstructing clinical activities; patients records in medical history", *Social History of Medicine*, 2 (1992), pp. 183–205.

Risse, Günter, "Hysteria at the Edinburgh Infirmary: The construction and treatment of a disease, 1770-1800", *Medical History*, 32 (1988), pp. 1–22.

Robcis, Camille, *Disalienation. Politics, Philosophy, and Radical Psychiatry in Postwar France*, Chicago: University of Chicago Press, 2021.

Rogers, Anne and David Pilgrim, *A Sociology of Mental Health and Illness*, 6th edition, London: Open University Press, 2021.

Rose, Nikolas, *Inventing our Selves: Psychology, Power, and Personhood*, Cambridge: Cambridge University Press, 1996.

Rothman, David, "Social control: The uses and abuses of the concept in the history of incarceration", in Stanley Cohen and Andrew Scull (eds), *Social Control and the State: Historical and Comparative Essays*, Oxford: Robertson, 1983 pp. 106–17.

Rothman, David, *The Discovery of the Asylum: Social Order and Disorder in the New Republic*, Boston: Little, Brown, 1971.

Rousseau, G.S., "'A strange pathology': Hysteria in the early modern world, 1500–1800", in Sander Gilman, Helen King, Roy Porter, G. S. Rousseau and Elaine Showalter (eds), *Hysteria Beyond Freud,* Berkeley: University of California Press, 1993, pp. 91–221.

Sadowsky, Jonathan, *Imperial Bedlam: Institutions of Madness in Colonial Southwest Nigeria*, Berkeley: University of California Press, 1999.

Salimpa, Zizi, *Γυναίκες Εργάτριες στην Ελληνική Βιομηχανία και στη Βιοτεχνία (1870–1922) [Women Workers in Greek Industry and Manufacture (1870–1922)]*, Athens: Ινστιτούτο Νεοελληνικών Ερευνών/ Εθνικό Ίδρυμα Ερευνών [Institute of Modern Hellenic Studies/ National Hellenic Research Foundation], 2002.

Schwieso, Joshua John, "Religious fanaticism and wrongful confinement in Victorian England", *Social History of Medicine,* 9 (1996), pp. 159–74.

Scourfield, Jonathan, "Suicidal masculinities", *Sociological Research Online*, 10(2), 2005.

Scull, Andrew, "The social history of psychiatry in the Victorian era", in Andrew Scull (ed.), *Madhouses, Mad-Doctors, and Madmen. The Social History of Psychiatry in the Victorian Era*, London: Athlone Press, 1981, pp. 5–32.

Scull, Andrew, Charlotte MacKenzie and Nicholas Hervey, *Masters of Bedlam: The Transformation of the Mad-Doctoring Trade*, Princeton, N.J.: Princeton University Press, 1996.

Scull, Andrew, *Madness in Civilization: A Cultural History of Insanity from the Bible to Freud, from the Madhouse to Modern Medicine*, London: Thames & Hudson, 2015.

Scull, Andrew, *Museums of Madness: The Social Organization of Insanity in Nineteenth-Century England*, London: Allen Lane, 1979.

Scull, Andrew, *Social Order/Mental Disorder: Anglo-American Psychiatry in Historical Perspective*, London: Routledge, 1989.

Scull, Andrew, *The Most Solitary of Afflictions: Madness and Society in Britain, 1700-1900*, New Haven: Yale University Press, 1993.

Sedgwick, Peter, *Psycho Politics,* London: Pluto, 1982.

Sengoopta, Chandak, "'A mob of incoherent symptoms'? Neurasthenia in British medical discourse, 1860-1920", in Marijke Gijswijt-Hofstra and Roy Porter (eds), *Cultures of Neurasthenia from Beard to the First World War*, Amsterdam: Rodopi, 2001, pp. 97–116.

Shephard, Ben, *A War of Nerves*, London: Jonathan Cape, 2000.

Shepherd, Anna, *Institutionalizing the Insane in Nineteenth-Century England*, London: Pickering & Chatto, 2014.

Shorter, Edward, *A Historical Dictionary of Psychiatry*, Oxford: Oxford University Press, 2005.

Shorter, Edward, *A History of Psychiatry: From the Era of the Asylum to the Age of Prozac*, New York: John Wiley, 1997.

Shorter, Edward, *From Paralysis to Fatigue: A History of Psychosomatic Illness in the Modern Era*, New York: Free Press, 1992.

Showalter, Elaine, "Hysteria, feminism and gender", in Sander Gilman, Helen King, Roy Porter, G. S. Rousseau and Elaine Showalter (eds), *Hysteria Beyond Freud*, Berkeley: University of California Press, 1993, pp. 286–344.

Showalter, Elaine, *The Female Malady, Women, Madness and English Culture, 1830-1980*, New York: Pantheon Books, 1985.

Sifnaiou, Evridiki, *Έλληνες Έμποροι στην Αζοφική. Η Δύναμη και τα Όρια της Οικογενειακής Επιχείρησης [Greek Merchants in the Sea of Azov. The Power and Limits of Family Enterprise]*, Athens: Ινστιτούτο Νεοελληνικών Ερευνών/ Εθνικό Ίδρυμα Ερευνών [Institute of Modern Hellenic Studies/ National Hellenic Research Foundation], 2009.

Smith-Rosenberg, Carroll, "The hysterical woman: Sex roles and role conflict in 19th-century America" (1972), in Carroll Smith-Rosenberg, *Disorderly Conduct: Visions of Gender in Victorian America*, New York: A.A. Knopf, 1985, pp. 197–216.

Snelders, Stephen, Frans J. Meijman and Toine Pieters, "Heredity and alcoholism in the medical sphere: The Netherlands, 1850-1900", *Medical History*, 51 (2007), pp. 219–36.

Sofianos, Kostas, *Το Νομικό Καθεστώς της Παιδικής Ηλικίας και της Νεότητας 1833–1900: Συναγωγή Νόμων, Διαταγμάτων, Εγκυκλίων, Καταστατικών [The Legal Status of Childhood and Youth 1833–1900: A Compilation of Laws, Decrees, Circulars, Statutes]*, Athens: Γενική Γραμματεία Νέας Γενιάς [General Secretariat for Youth], 1988.

Still, Arthur and Irving Velody, *Rewriting the History of Madness: Studies in Foucault's Histoire de la Folie*, London: Routledge, 1992.

Strange, Julie-Marie, "Menstrual fictions: Languages of medicine and menstruation, c. 1850-1930", *Women's History Review*, 9 (2000), pp. 607–28.

Strange, Julie-Marie, "The assault on ignorance: Teaching menstrual etiquette in England, c. 1920s to 1960s", *Social History of Medicine*, 14 (2001), pp. 247–65.

Suzuki, Akihito, "Were asylums men's places?: Male excess in the asylum population in Japan in the early twentieth century", in Marijke Gijswijt-Hofstra, Harry Oosterhuis, Joost Vijselaar and Hugh Freeman (eds), *Psychiatric Cultures Compared: Psychiatry and Mental Health Care in the Twentieth Century. Comparisons and Approaches*, Amsterdam: Amsterdam University Press, 2006, pp. 295–311.

Suzuki, Akihito, *Madness at Home: The Psychiatrist, the Patient, and the Family in England, 1820-1860*, Berkeley: University of California Press, 2006.

Swain, Gladys, *Le Sujet de la Folie: Naissance de la Psychiatrie*, Toulouse: Privat, 1977.

Swartz, Sally, "Mad Africa", in Greg Eghigian (ed.), *The Routledge History of Madness and Mental Health*, London: Routledge, 2017, pp. 229–44.

Szasz, Thomas, *The Myth of Mental Illness: Foundations of a Theory of Personal Conduct*, New York: Harper & Row, 1974.

Thane, Pat, *Old Age in English History: Past Experiences, Present Issues*, Oxford: Oxford University Press, 2000.

Theodorou, Vasiliki, "Αλητεία και επαιτεία στην Αθήνα και τον Πειραιά. Οι πρώτες απόπειρες 'αναμόρφωσης παραστρατημένων παίδων' (τέλη 19ου αρχές 20ού αιώνα)" ["Truancy and begging in Athens and Piraeus. The first attempts to 'reform misguided children' (late 19th and early 20th century"], in *Η Πόλη στους Νεότερους Χρόνους: Μεσογειακές και Βαλκανικές Όψεις (19ος-20ός Αιώνας) [The City in Modern Times: Mediterranean and Balkan Aspects (19th-20th Century)]*, Proceedings of the 2nd International Conference of History, Athens: Εταιρεία Μελέτης Νέου Ελληνισμού [Society for the Study of Modern Hellenism], 2000, pp. 203–30.

Theodorou, Vasiliki, "Μεταβαλλόμενα πλαίσια συνάφειας μεταξύ εθελοντών, ειδικών και κράτους: Το παράδειγμα του Πατριωτικού Ιδρύματος Προστασίας του Παιδιού" ["Changing contexts of affinity between volunteers, experts and the state: The example of the Patriotic Foundation for the Protection of Children"], in Efi Avdela, Haris Exertzoglou, Christos Lyrintzis (eds), *Μορφές Δημόσιας Κοινωνικότητας στην Ελλάδα του Εικοστού Αιώνα [Forms of Public Sociality in Twentieth Century Greece]*, Rethymno: Πανεπιστήμιο Κρήτης [University of Crete], 2015, pp. 82–100.

Theodorou, Vasiliki, "Φιλανθρωπία και πόλη. Ορφανοί και άστεγοι παίδες στον Πειραιά γύρω στο 1875" ["Charity and the city. Orphans and homeless children in Piraeus around 1875"], *Μνήμων [Mnimon]*, 14 (1992), pp. 71–90.

Theodorou, Vassiliki and Despina Karakatsani, "Eugenic concerns, population policies and puericulture in interwar Greece", *The Historical Review/La revue Historique*, 17 (2020), pp. 53–90.

Theodorou, Vassiliki, "La Fondation patriotique pour la protection de l'enfant en Grèce. De l'action des philanthropes à l'organisation des services publics de santé et de prévoyance pour la maternité et l'enfance (1914-1940)", *Revue d'Histoire de la Protection Sociale*, 11:1 (2018), pp. 72–91.

Thifault, Marie-Claude and Isabelle Perreault, "Behind asylum walls: Studying the dialectic between psychiatrists and patients at Montreal's Saint-Jean-de-Dieu Hospital during the first half of the twentieth century", *Social History of Medicine*, 32 (2019), pp. 377–94.

Thomson, Margaret, "The wages of sin: The problem of alcoholism and general paralysis in nineteenth-century Edinburgh", in William Frederick Bynum, Roy

Porter and Michael Shepherd (eds), *The Anatomy of Madness: Essays in the History of Psychiatry,* v. 3, London: Tavistock, 1988, pp. 316–40.

Thomson, Mathew, "Neurasthenia in Britain: An overview", in Marijke Gijswijt-Hofstra and Roy Porter (eds), *Cultures of Neurasthenia from Beard to the First World War*, Amsterdam: Rodopi, 2001, pp. 77–96.

Todd, John and Lawrence Ashworth, "The West Riding Asylum and James Crichton-Browne, 1818-1876", in German Berrios and Hugh Freeman (eds), *150 Years of British Psychiatry, 1841-1991*, London: Gaskell, 1991, pp. 389–418.

Tomes, Nancy, "A generous confidence: Thomas Story Kirkbride's philosophy of asylum construction and management", in Andrew Scull (ed.), *Madhouses, Mad-doctors, and Madmen: The Social History of Psychiatry in the Victorian Era*, London: Athlone Press, 1981, pp. 120–143.

Tomes, Nancy, "The anatomy of madness: New directions in the history of psychiatry", *Social Studies of Science*, 17 (1987), pp. 358–70.

Tomes, Nancy, *The Art of Asylum-Keeping: Thomas Story Kirkbride and the Origins of American Psychiatry*, Philadelphia: University of Pennsylvania Press, 1994.

Trede, K. et al, "Manic depressive illness: Evolution in Kraepelin's textbook, 1883-1926", *Harvard Review of Psychiatry*, 13 (2005), pp. 155–78.

Trubeta, Sevasti, *Physical Anthropology, Race and Eugenics in Greece 1880s-1970s*, Boston: Brill Academic Publishers, 2013.

Tsichli-Aroni, Keti, "Το αγροτικό ζήτημα. Θεσσαλία- Νέες Χώρες, 1909–1922" ["The agricultural issue. Thessaly - New Territories, 1909–1922"], in Vasilis Panayotopoulos (ed.), *Ιστορία του Νέου Ελληνισμού, 1770–2000 [History of Modern Hellenism, 1770–2000]*, v. 6, Athens: Ελληνικά Γράμματα [Ellinika Grammata], 2003, pp. 267–82.

Tsikis, Nikolaos (ed.), *Δρομοκαΐτειο Θεραπευτήριο, 1887–1987+15. Εκατό Δεκαπέντε Χρόνια Κοινωνικής Προσφοράς [Dromokaitio Sanatorium, 1887–1987+15. One Hundred Fifteen Years of Social Offer]*, Athens: s.n., 2001.

Turner, Trevor, "A diagnostic analysis of the casebooks of Ticehurst House Asylum, 1845–1890", *Psychological Medicine Supplements*, 21 (1992), pp. 1–70.

Tzanakis, Manolis, "Μετασχηματισμοί υποκειμενικότητας και καθημερινότητα. Μια κοινωνιολογική προσέγγιση της ψυχιατρικής μεταρρύθμισης στην Ελλάδα" ["Transformations of subjectivity and everyday life. A sociological approach of psychiatric reform in Greece"], PhD thesis, University of Crete, 2003.

Tzavaras, Nikos, "Κοινωνιολογικές και ψυχιατρικές διαστάσεις του αλκοολισμού" ["Sociological and psychiatric dimensions of alcoholism"], *Σύναψις [Synapsis]*, 1 (2005), pp. 64–72.

Valverde, Mariana, *Diseases of the Will: Alcohol and the Dilemmas of Freedom*, Cambridge: Cambridge University Press, 1998.

Varika, Eleni, *Η Εξέγερση των Κυριών. Η Γένεση μιας Φεμινιστικής Συνείδησης στην Ελλάδα, 1933–1907 [The Rising of the Ladies. The Genesis of a Feminist Consciousness in Greece, 1933–1907]*, Athens: Κατάρτι [Katarti], 2004.

Vassiliadou, Dimitra, *Στον Τροπικό της Γραφής. Οικογενειακοί Δεσμοί και Συναισθήματα στην Αστική Ελλάδα, 1850–1930 [In the Tropic of Writing. Family Ties and Emotions in Urban Greece, 1850–1930]*, Athens: Gutenberg, 2018.

Vatz, Richard and Lee Weinberg, "The rhetorical paradigm in psychiatric history: Thomas Szasz and the myth of mental illness", in Mark Micale and Roy Porter (eds), *Discovering the History of Psychiatry*, New York: Oxford University Press, 1994, pp. 311–30.

Vijselaar, Joost, "Out and in: The family and the asylum. Patterns of admission and discharge in three Dutch psychiatric hospitals 1890–1590", in Marijke Gijswijt-Hofstra, Harry Oosterhuis, Joost Vijselaar and Hugh Freeman (eds), *Psychiatric Cultures Compared: Psychiatry and Mental Health Care in the Twentieth Century: Comparisons and Approaches*, Amsterdam: Amsterdam University Press, 2006, pp. 277–94.

Vreugdenhil, Anthea, "'Incoherent and violent if crossed': The admission of older people to the New Norfolk Lunatic Asylum in the nineteenth century", *Health and History*, 14 (2012), pp. 91–111.

Wallis, Alexandra, "Driven to insanity: Marital cruelty and the female patients at the Fremantle Lunatic Asylum, 1858-1908", *Limina*, 24 (2019), pp. 1–16.

Wallis, Alexandra, "Hysterical women: Moral treatment of female patients in the Fremantle Lunatic Asylum, 1858–1908", PhD thesis, The University of Notre Dame Australia, 2020.

Wallis, Jennifer, *Investigating the Body in the Victorian Asylum: Doctors, Patients, and Practices*, London: Palgrave Macmillan, 2017.

Walmsley, Jan, "Mental defectives in 1900: People with learning disabilities in 2000. What changed?: A response to Matt Egan's paper", in *Scottish Health History: International Contexts, Contemporary Perspectives*, Glasgow: Centre for the History of Medicine, University of Glasgow, 2003.

Walsh, Lorraine, "A class apart? Admissions to the Dundee Royal Lunatic Asylum, 1890-1910", in Jonathan Andrews and Anne Digby (eds), *Sex and Seclusion, Class and Custody: Perspectives on Gender and Class in the History of British and Irish Psychiatry*, Amsterdam: Rodopi, 2004, pp. 249–69.

Walton, John, "Casting out and bringing back in Victorian England: Pauper lunatics, 1840–70", in William Frederick Bynum, Roy Porter and Michael Shepherd (eds), *The Anatomy of Madness: Essays in the History of Psychiatry*, v. 2, London: Tavistock, 1985, pp. 132–46.

Waltraud, Ernst, "Histories of madness in South Asia", in Greg Eghigian (ed.), *The Routledge History of Madness and Mental Health*, London: Routledge, 2017, pp. 210–28.

Wannell, Louise, “Patients’ relatives and psychiatric doctors: Letter writing in the York Retreat, 1875-1910”, *Social History of Medicine* 20 (2007), pp. 297–313.

Wilde, Sally, “The elephants in the doctor-patient relationship: Patients’ clinical interactions and the changing surgical landscape of the 1890s”, *Health and History*, 9 (2007), pp. 2–27.

Winter, Jay, “Shell-shock and the cultural history of the Great War”, *Journal of Contemporary History*, 35 (2000), pp. 7–11.

Wright David, *Mental Disability in Victorian England: The Earlswood Asylum, 1847-1901*, Oxford: Clarendon Press, 2001.

Wright, David, “‘Childlike in his innocence’. Lay attitudes to ‘idiots’ and ‘imbeciles’ in Victorian England”, in David Wright and Anne Digby (eds), *From Idiocy to Mental Deficiency. Historical Perspectives on People with Learning Disabilities*, London: Routledge, 1996, pp. 118–33.

Wright, David, “Delusions of gender? Lay identification and clinical diagnosis of insanity in Victorian England”, in Jonathan Andrews and Anne Digby (eds), *Sex and Seclusion, Class and Custody: Perspectives on Gender and Class in the History of British and Irish Psychiatry*, Amsterdam: Rodopi, 2004, pp. 149–76.

Wright, David, “Getting out of the asylum: Understanding the confinement of the insane in the nineteenth century”, *Social History of Medicine*, 10 (1997), pp. 137–55.

Yannitsiotis, Yannis, *Η Κοινωνική Ιστορία του Πειραιά, 1860–1910. Η Συγκρότηση της Αστικής Τάξης [The Social History of Piraeus, 1860–1910. The Formation of the Urban Class]*, Athens: Νεφέλη [Nefeli], 2006.

Yanniris, Michalis, “Η επιστημονική δραστηριότητα του Δρομοκαΐτειου στα πρώτα πενήντα χρόνια της λειτουργίας του” [“The scientific activity of the Dromokaiteio in the first fifty years of its operation”], in Nikolaos Tsikis (ed.), *Δρομοκαΐτειο Θεραπευτήριο, 1887–1987+15. Εκατό Δεκαπέντε Χρόνια Κοινωνικής Προσφοράς [Dromokaitio Sanatorium, 1887–1987+15. One Hundred Fifteen Years of Social Offer]*, Athens: s.n., 2001, pp. 95–103.

Yi-Jui, Wu Harry, *Mad by the Millions: Mental Disorders and the Early Years of the World Health Organization*, Cambridge, MA: MIT Press, 2020.

Yvorel, Jean-Jacques, “Les mots pour le dire: Naissance du concept de toxicomanie”, in Francois-Olivier Touati (ed.), *Maladies, Médecines et Sociétés: Approches Historiques pour le Present.* Actes du VIe Colloque, v. 1, Paris: L’Harmattan, 1993, pp. 209–17.

Ziff, Katherine, David Thomas and Patricia Beamish, “Asylum and community: The Athens Lunatic Asylum in nineteenth-century Ohio”, *History of Psychiatry*, 19 (2008), pp. 409–32.

Zilboorg, Gregory and George Henry, *A History of Medical Psychology*, New York: W. W. Norton & Company, 1941.

Ziogou-Karastergiou, Sidiroula, *Γυναίκες και Ανώτατη Εκπαίδευση στην Ελλάδα. Οι Πρώτες Φοιτήτριες στο Πανεπιστήμιο Αθηνών (1890–1920) [Women and Higher Education in Greece. The First Female Students at the University of Athens (1890–1920)]*, Thessaloniki: n.p., 1988.

Index

D. Kritsotaki, *A Social History of Mental Health Care in Greece, c. 1900–1920*, Mental Health in Historical Perspective,
https://doi.org/10.1007/978-3-032-05764-8

Kolibri 360 GmbH
Ferdinand-Jühlke-Straße 7
99095 Erfurt, Deutschland
produktsicherheit@kolibri360.de